Praise for *Future Perfect*

"*Future Perfect* provides an informative, tech-savvy, and provocative vision of a new and more democratic public philosophy. It's a breath of fresh air in an age of gridlock, cynicism, and disillusionment." —*San Francisco Chronicle*

"Mr. Johnson envisions a new political movement that embraces the potential of peer networks to improve government, medicine, education, and journalism, among much else. He distinguishes 'peer progressives' from both libertarians and liberals. The former have too much faith in markets and too little in government, he says, and the latter vice versa. Peer progressives, though, believe that good can be accomplished by all organizations, in any combination, if they harness the power of peer networks." —*The Wall Street Journal*

"In clear and engaging prose, Johnson writes about this emerging movement. . . . *Future Perfect* is a buoyant and hopeful book. Given the inability of our government to enact worthwhile change, and the near guarantee that Washington's gridlock will only worsen regardless of which party wins this November, we're going to need all the help we can get. *Future Perfect* reminds us we already have the treatment. We just need to use it." —*The Boston Globe*

"Forceful argument for a new politics modeled on the structure of the Internet. . . . A thought-provoking, hope-inspiring manifesto." —*Kirkus Reviews*

"A wide-ranging sketch of possibilities . . . frequently inspiring. Above all, it's exciting to reflect on the possibility that the many achievements of the Silicon Valley revolution might be compatible, rather than in tension, with a progressive focus on social justice and participatory democracy." —*The Guardian* (UK)

"Fascinating and compelling . . . Stimulating and challenging, Johnson's thought-provoking ideas steer us steadily into the future." —*Publishers Weekly*

Future Perfect

The Case for Progress in a Networked Age

Steven Johnson

RIVERHEAD BOOKS

New York

RIVERHEAD BOOKS
Published by the Penguin Group
Penguin Group (USA)
375 Hudson Street, New York, New York 10014, USA

USA | Canada | UK | Ireland | Australia | New Zealand | India | South Africa | China

Penguin Books Ltd., Registered Offices: 80 Strand, London WC2R oRL, England
For more information about the Penguin Group, visit penguin.com.

The Library of Congress has catalogued the Riverhead hardcover edition as follows:

Johnson, Steven, date.
Future perfect : the case for progress in a networked age / Steven Johnson.
p.   cm.
Includes bibliographical references and index.
ISBN 978-1-59448-820-7
1. Information technology—Social aspects.   2. Progress.   3. Information networks.
4. Social networks.   I. Title.
HM851.J63   2012   2012026086
303.48'33—dcsd

First Riverhead hardcover edition: September 2012
First Riverhead trade paperback edition: October 2013
Riverhead trade paperback ISBN: 978-1-59463-184-9

PRINTED IN THE UNITED STATES OF AMERICA

10   9   8   7   6   5   4   3   2   1

Cover design by Alex Merto
Book design by Chris Welch

For Clay, Rowan, and Dean

Contents

Introduction

PROGRESS, ACTUALLY

Commercial airplane crashes are one of the few events guaranteed a place on the front page of every national newspaper. Somewhat less common on the front page, however, is a story about a commercial airplane crash that didn't happen. But every now and then one of these non-events manages to cross the threshold into newsworthiness. One such unlikely story appeared on page one of the January 12, 2009, edition of *USA Today*: "Airlines Go Two Years with No Fatalities," the headline proclaimed. An analysis by the paper had discovered that the U.S. commercial aviation industry had achieved a milestone unprecedented in the history of modern air travel: despite the dramatic increase in flights compared with earlier decades, not a single person had perished in a commercial air accident during the years 2007 and 2008.

A two-year stretch with zero fatalities was, statistically speaking, a remarkable feat. Only four times since 1958 had the industry managed to make it through one year without a fatal accident. But the safety record was part of a long-term trend: in the post-9/11 period, chances of dying on a commercial flight were nineteen in 1 billion; an almost 100 percent improvement over the already excellent odds of flying in the 1990s. According to MIT professor Arnold Barnett, this meant that an American child was more likely to be elected president of the United States in his or her lifetime than to die in a commercial jet flight.

The story caught my eye in part because I had long maintained that news organizations presented a misleading slant on current events—not the conventional slant of the Left-Right political world, but the more subtle bias of being more interested in bad news. "If it bleeds, it leads" might be a good strategy for selling papers, but it necessarily skews our collective sense of how well we are doing as a society. We hear about every threat or catastrophe, but the stories of genuine progress get relegated to the back pages, if they run at all.

But this *USA Today* article was bucking that trend of negative bias. And so, inspired by its refreshing story of progress, I decided to write a little essay about it, for a website at which I happened to be a guest contributor that week. I summarized the article's findings, and made the point that we might have fewer people suffering from fear of flying (and thus spending less time on much more dangerous highways) if the media did a better job of reminding us of these genuinely extraordinary achievements.

A more superstitious person might have worried that saying such things in public was tempting fate. In fact, when I had talked about airline safety with friends in the past, I'd often been accused of just that, a criticism that I would inevitably laugh off. Surely fate had better things to do than worry about my monologues.

But within a few hours of publishing my little riff about airline safety, while I was sitting in a television studio waiting to do a book promotion, my phone began buzzing with e-mails and text messages from friends who had seen my piece. A US Airways jet had crash-landed in the icy Hudson. Look what you did, my friends said. I had tempted fate, and fate had slapped me back across the face.

By the time I actually managed to get to a television set to see the live report, it was clear that the plane had in fact survived the landing with minimal structural damage, and a significant number of survivors were clearly visible standing on the wings, waiting for the ferries and police boats that would rescue them. I penned a quick, sheepish entry promising to keep mum about air safety for a while, and mothball that piece I was working on about the planet's amazing streak of avoiding species-extinguishing asteroid collisions.

Then a funny—and wonderful—thing happened. The world quickly learned that all 155 passengers and crew aboard flight 1549 had survived, with almost no serious injuries. Flight 1549 went from being a devastating rebuke to my optimism about commercial aviation to being a ringing endorsement of it. If you believed in that sort of thing, you might have said that I had tempted fate, and fate had blinked.

Of course, I had nothing to do with the safe landing of US Airways 1549. The question was, Who or what was responsible for that plane managing to lose two engines during takeoff and still keep its cargo secure? The mass media quickly offered up two primary explanations, both of which turned out to be typical interpretations of good news. First there was the hero narrative: Captain Chesley "Sully" Sullenberger, who had indeed brilliantly navigated his plane into the river with great poise under unthinkable pressure. And then there was the quasi-magical rhetoric that quickly became attached to the event, the Miracle on the Hudson. Those were the two options. That plane floating safely in the Hudson could be explained only by superheroes or miracles.

There was no denying Sullenberger's achievement that day, but the fact is, he was supported by a long history of decisions made by thousands of people over the preceding decades, all of which set up the conditions that made that perfect landing possible. A lesser pilot could have still failed catastrophically in that situation, but as good as Sullenberger was, he was not working alone. It was surprising and thrilling that all 155 people survived the crash, but there was nothing miraculous about it. The plane survived because a dense network of human intelligence had built a plane designed to withstand exactly this kind of failure. It was an individual triumph, to be sure, but it was also, crucially, a triumph of collectively shared ideas, corporate innovation, state-funded research, and government regulation. To ignore those elements in telling the story of the Miracle on the Hudson is not to neglect part of the narrative for dramatic effect. It is to fundamentally misunderstand where progress comes from, and how we can create more of it.

Any attempt to explain the confluence of events that came together to allow flight 1549 to land safely in the Hudson has to begin with the chicken gun.

The threat posed by bird-impact strikes to aircraft dates back to the very beginnings of flight. (The Wright brothers recorded bird strikes in their journals during their experimental flights in the early 1900s.) The primary vulnerability in a modern commercial jet lies in birds' being ingested by the jet engine, and wreaking enough internal damage that the engine itself fails. But there are degrees of failure. An engine can simply flame out and stop functioning. Or it can shatter, sending debris back into the fuselage, potentially destroying the plane in a matter of seconds. The former, it goes without saying, is greatly preferable to the latter. Most bird strikes involve only one engine, leaving the plane perfectly capable of flying on its remaining engines if the main structure of the plane is still intact.

Today's jet engines are therefore rigorously tested to ensure that they can withstand significant bird impact without catastrophic failure. At Arnold Air Force Base in Tennessee, a team of scientists and engineers use high-pressure helium gas to launch chicken carcasses at high velocity into spinning jet engines. Every make of engine that powers a commercial jet aircraft in the United States has passed the chicken gun test.

The chicken gun, it should be noted, is an exemplary case of government regulation. Those dead birds being shot out of a pneumatic cannon are Your Tax Dollars at Work. For the passengers flying on US Airways 1549, those tax dollars turned out to be very well spent. Because the first lucky break the plane experienced after a flock of Canada geese crashed into both its engines was the simple fact that neither engine disintegrated. Neither one propelled shards of titanium into the fuselage; neither engine caught fire.

The phrase "lucky break"—like the whole premise of a Miracle on the Hudson—distorts the true circumstances of the US Airways landing. We need a better phrase, something that conveys the idea of an event that seems lucky, but actually resulted from years of deliberate preparation and planning. This was not a stroke of good fortune. It was a stroke of good foresight.

In fact, the advance planning of the chicken gun was so effective that the jet core of the left engines continued to spin at near-maximum speed—not enough to grant Sullenberger the thrust he needed to return to LaGuardia, but enough so that the plane's electronics and hydraulic systems functioned for the duration of the flight. The persistence of the electronics system, in turn, set up flight 1549's second stroke of foresight: the plane's legendary fly-by-wire system remained online as Sullenberger steered his wounded craft toward the river.

The history of fly-by-wire dates back to 1972, when a modified F-8 Crusader took off from the Dryden Flight Research Center on the edge of the Mojave Desert. The brainchild of NASA engineers, the fly-by-wire system used digital computers and other modern electronic systems to relay control information from the pilot to the plane. Because computers were involved, it became easier to provide assistance to the pilot in real time, even if the autopilot was disengaged, preventing stalls, or stabilizing the plane during turbulence. Inspired by the NASA model, engineers at Airbus in the early 1980s built an exceptionally innovative fly-by-wire system into the Airbus A320, which began flying in 1987.

Twenty-one years later, Chesley Sullenberger was at the controls of an A320 when he collided with that flock of Canada geese. Because his left engine was still able to keep the electronics running, his courageous descent into the Hudson was deftly assisted by a silent partner, a computer embodied with the collective intelligence of years of research and planning. William Langewiesche describes that digital aid in his riveting account of the flight, *Fly by Wire*:

> While in the initial left turn [Sullenberger] lowered the nose . . . and went to the best gliding speed—a value which the airplane calculated all by itself, and presented to him as a green dot on the speed scale of his primary flight display. During the pitch changes to achieve that speed, a yellow "trend" arrow appeared on the scale, pointing up or down from the current speed with predictions of speed 10 seconds into the future—an enormous aid in settling onto the green dot with the minimum of oscillation. . . .

Whenever he left the side-stick alone in the neutral position the airplane held its nose steadily at whatever pitch he had last selected; that the airplane's pitch trim was automatic, and perfect at all times.

Most non-pilots think of modern planes as possessing two primary modes: "autopilot," during which the computers are effectively flying the plane, and "manual," during which humans are in charge. But fly-by-wire is a more subtle innovation. Sullenberger was in command of the aircraft as he steered it toward the Hudson, but the fly-by-wire system was silently working alongside him throughout, setting the boundaries or optimal targets for his actions. That extraordinary landing was a kind of duet between a single human being at the helm of the aircraft and the embedded knowledge of the thousands of human beings that had collaborated over the years to build the Airbus A320's fly-by-wire technology. It is an open question whether Sullenberger would have been able to land the plane safely without all that additional knowledge at his service. But fortunately for the passengers of flight 1549, they didn't have to answer that question.

The popular response to the Miracle on the Hudson encapsulates just about everything that is flawed in the way we think about progress in our society. First, the anomalous crash landing (fatal or not) gets far more play than the ultimately more important story of long-term safety trends. As a news hook, steady, incremental progress pales beside the sexier stories of dramatic breakthrough and spectacular failure. For reasons that are interesting to explore, it also pales beside stories of steady, incremental decline. You can always get bandwidth by declaring yourself a utopian; and you can always get bandwidth by mourning the downward trend lines for some pressing social issue—however modest the trend itself may be. But declaring that things are slightly better than they were a year ago, as they have been for most years since at least the dawn of industrialization, almost never makes the front page.

Consider this observation from the entrepreneur and investor Peter Thiel, published in *National Review*:

> When tracked against the admittedly lofty hopes of the 1950s and 1960s, technological progress has fallen short in many domains. Consider the most literal instance of non-acceleration: We are no longer moving faster. The centuries-long acceleration of travel speeds—from ever-faster sailing ships in the 16th through 18th centuries, to the advent of ever-faster railroads in the 19th century, and ever-faster cars and

airplanes in the 20th century—reversed with the decommissioning of the Concorde in 2003, to say nothing of the nightmarish delays caused by strikingly low-tech post-9/11 airport-security systems. Today's advocates of space jets, lunar vacations, and the manned exploration of the solar system appear to hail from another planet. A faded 1964 *Popular Science* cover story—"Who'll Fly You at 2,000 m.p.h.?"— barely recalls the dreams of a bygone age.

But raw airspeed is only one unit by which we can measure our transportation progress. It happens to be the sexiest metric, the one that gets the headlines when the first commercial jets hit the skies, or the Concorde breaks the sound barrier. But focusing on that metric alone distorts our broader measurement of progress in modern aviation. Most passengers would probably value safety over speed, particularly if the speed already on the table happens to be 600 mph. And indeed, our progress in aviation safety is off the charts. One small historical anecdote illustrates how far we have progressed: In many airports in 1964, a few steps down from that copy of *Popular Science* at the newsstand, you could find a little booth where you could make a last-minute purchase of life insurance right before you boarded the flight. Not exactly an encouraging sight, but that foreboding was appropriate. At the time, your odds of dying in a commercial jet crash were roughly one in a million. Today it is a hundred times safer. If jet velocity had increased at that same pace, flying from New York to Paris would now take about five minutes.

Even if the raw airspeed of commercial jets has flatlined over the past forty years, average travel times have none-theless decreased, because it is now so much easier to fly to midsized markets, thanks to the growth of the overall in-dustry and the modern hub system that creates a flexible network of large jets and smaller regional planes that de-posit passengers on the less populated spokes of the net-work. If you were flying from New York to Los Angeles in 1970, you'd get there about as quickly as you would today. But if you were flying from New York to, say, Jackson Hole, Wyoming, the trip might take days, not hours. The engines aren't faster, but the overall transportation system is. That's progress, too.

And then there's price. The New York–L. A. round trip would have cost you roughly $3,000 to fly coach, in inflation-adjusted dollars. Today, you can easily find a flight for the same itinerary for $500, and watch live satellite tele-vision or check your e-mail as you fly.

Yes, Thiel is right that the planes themselves can't fly any faster than they did forty years ago, and so by that metric, progress has in fact stalled. (Or gone backward, if you count the Concorde.) But just about every other crucial metric (other than the joys of going through airport security) points in the other direction. That extraordinary record of progress did not come from a breakthrough device or a visionary inventor; it did not take the form of a great leap forward. Instead, the changes came from decades of small decisions, made by thousands of individuals and organizations, some of them public-sector and some of them private, each tinkering with the system in tactical ways: exploring new routes, experimenting with new pricing structures, throwing chicken carcasses into spinning jet engines. Each of these changes was incremental, but over time they built themselves up into orders-of-magnitude improvements. Yet because they were incremental, they remained largely invisible, unsung.

If modern aviation history isn't enough to convince you that we are biased against incremental progress, then take this brief social studies quiz. Over the past two decades, what have the U.S. trends been for the following important measures of social health: high school dropout rates; college enrollment; SAT scores; juvenile crime; drunk driving; traffic deaths; infant mortality; life expectancy; per capita gasoline consumption; workplace injuries; air pollution; divorce; male-female wage equality; charitable giving; voter turnout; per capita GDP; and teen pregnancy?

The answer for all of them is the same: The trend is positive. The progress is not as dramatic as the story of airline safety over that period, but almost all those varied metrics of social wellness have improved by more than 20 percent over the past two decades. (See the Notes section for a longer discussion of each trend.) Add to that the myriad small wonders of modern medicine that have improved our quality of life as well as our longevity: the antidepressants and insulin pumps and quadruple bypasses. (Not to mention the panoply of entertainment and communications devices whose prices have plummeted during that period.) Americans enjoy a longer, healthier life in more stable families and communities than they did twenty years ago, surrounded by an array of amusing and laborsaving technologies that exceed anything you would have found in a palace a century ago. But other than the crime trends and the gadgets, these facts are rarely reported or shared via word-of-mouth channels. Many Americans, for instance, are convinced that "half of all marriages end in divorce," though that hasn't been the case since the early 1980s, when divorce rates peaked at just over 50 percent. Since then, they have declined by almost a third.

This is not merely a story of success in advanced industrial countries. The quality-of-life and civic health trends in the developing world are even more dramatic. Even though the world's population has doubled over the past fifty years, the percentage living in poverty has declined by 50 percent over that period. Infant mortality and life expectancy have improved by more than 40 percent in Latin America since the early 1990s. No country in history has improved its average standard of living faster than China has over the past two decades.

Of course, not all the arrows point in a positive direction, particularly after the last few years. The number of Americans living in poverty has increased over the last decade, after a long period of decline. Wealth inequality has returned to levels last seen in the Roaring Twenties. As I write these words in early 2012, the U.S. unemployment rate is still more than 8 percent, higher than its average over the past two decades. Household debt has soared over the past twenty years, though it has dipped slightly, thanks to the credit crunch of the last few years. While most Americans are significantly healthier than they were a generation ago, childhood obesity has emerged as a meaningful problem, particularly in lower-income communities.

An interesting divide separates these two macro-trends. On the one hand, there is a series of societal trends that are heavily dependent on non-market forces. The progress made in preventing drunk driving or teen pregnancy or juvenile crime isn't coming from new gadgets or Silicon Valley start-ups or massive corporations; the progress, instead, is coming from a network of forces largely outside the marketplace: from government intervention, public service announcements, demographic changes, and the wisdom of life experience shared across generations. Capitalism didn't reduce the number of teen smokers; in fact, certain corporations did just about everything they could to keep those kids smoking (remember Joe Camel?). The decline in teen smoking came from doctors, regulators, parents, and peers sharing vital information about the health risks of smoking.

We don't hear enough about this kind of social progress for several reasons: First, we tend to assume that innovation and progress come from market environments, not the public sector. This propensity is no accident; it is the specific outcome of the way public opinion is shaped within the current media landscape. The public sector doesn't have billions of dollars to spend on marketing campaigns to trumpet its successes. If a multinational corporation invents a slightly better detergent, it will spend a legitimate fortune to alert the world that its product is now "new and improved." But no one launches a prime-time ad campaign to tout the chicken gun. The vast majority of public-sector dollars spent on advertising go into electing politicians. No one buys airtime to sing the praises of the regulators and the civil servants, so we assume the regulators and the civil servants have done nothing for us. The end result is a blind spot for stories of public-sector progress.

That blind spot is compounded by the deeper lack of interest in stories of incremental progress. Curmudgeons, doomsayers, utopians, and declinists all have an easier time getting our attention than opinion leaders who want to celebrate slow and steady improvement. The most striking example of this can be seen in the second half of the 1990s, a period when both economic and social trends were decisively upbeat: the stock market was surging, but inequality was in fact on the decline; crime, drug use, welfare dependence, poverty—all were trending in an encouraging direction. With a Democrat in the White House, you might assume that the op-ed pages of *The Washington Post* would be bursting with pride over the state of the nation, given the paper's center-left leanings. But you would be wrong. Over the course of 1997, in the middle of the greatest peacetime economic boom in U.S. history (and before the Monica Lewinsky scandal broke), 71 percent of all editorials published in the *Post* that expressed an opinion on some aspect of the country's current state focused on a negative trend. Less than 5 percent of the total number of editorials concentrated on a positive development. Even the boom years are a bummer.

I suspect, in the long run, the media bias against stories

of incremental progress may be more damaging than any bias the media display toward the political Left or Right. The media are heavily biased toward extreme events, and are slightly biased toward negative news and trend stories. This bias may just be a reflection of the human brain's propensity to focus more on negative information than positive, a trait extensively documented by neuroscience and psychology studies. The one positive social trend that did generate a significant amount of coverage—the extraordinary drop in the U.S. crime rate since the mid-1990s—seems to have been roundly ignored by the general public. The violent crime rate dropped from 51 to 15 (in crimes per thousand people) between 1995 and 2010, truly one of the most inspiring stories of societal progress in our lifetime. Yet according to a series of Gallup polls conducted over the past ten years, more than two-thirds of Americans believe that crime has been getting worse, year after year.

Whether these biases come from media distortions or our human psychology, they result in two fundamental errors in the popular mind: we underestimate the amount of steady progress that continues around us, and we misunderstand where that progress comes from.

In the American tradition, the word "progress" has long been embedded in one of the country's most durable political labels, dating back to the Progressive movement, which peaked a century ago with Teddy Roosevelt's failed presidential bid under the banner of the Progressive Party—to this day the most successful third-party challenge to the presidency since the modern two-party system consolidated in the middle of the nineteenth century. The original Progressives were inspired by two emerging developments. They shared a newfound belief in the importance of social justice for women and the working poor, embodied in the suffrage movement and the muckraking journalism that exposed the horrors of many industrial workplaces. And they shared a belief in a new kind of institution: the crusading Big Government that could use its power to combat the excesses of the capitalist oligarchs, by breaking up the monopolies, by supporting unions, by regulating conditions on the factory floor, and through other novel interventions.

The term "progressive" has had a revival in the past twenty years, in part because the word "liberal" was so successfully vilified by the political Right, and in part because self-identifying progressives wanted to distinguish themselves from the more moderate wing of the Democratic Party. But there was a funny thing about this recent generation of progressives: they didn't talk all that much about progress. If you gathered all the political persuasions in a room and got them talking, it would be the progressives who were most likely to talk doomsday scenarios around climate change, or the population bomb. Other persuasions had their complaints, of course: taxes were too high, or essential government services were being cut. But it was the progressives who seemed most likely to think that the human race was on an inexorable path toward self-annihilation.

The progressives' ambivalence about actual progress always struck me as a little odd. I had been drawn to the moniker because of its root. I wanted to call myself a progressive because I believed in progress. I believed that, on the whole, my generation was better off than my parents' generation, and that they had been better off than their parents' generation, and I believed that march upward would continue if we played our cards right. What's more, I liked talking about progress; I liked reminding people of all the things that we take for granted now: a life expectancy double that of a century ago; drinking water that didn't kill us; the creation of new tools for collaboration and civic participation. I liked talking about progress not because I thought we could rest on our laurels, but because talking about progress was a particularly effective way to inspire people. Life, on the whole, was getting better, and it had been for a while now—so why not gather together and dream up new ways to keep those trends going? That was the progressive tradition I wanted to belong to: one that was predicated on hope and optimism, not because those were nice slogans for a political campaign, but because there was plenty of evidence out there that suggested optimism was warranted.

It turned out that I was not alone. Sometime in the first, dark years after 9/11, I began to realize that a diverse group of people around me—some of them writers or academics, some of them entrepreneurs, some of them activists, some of them programmers, some of them even politicians—were starting to talk about progress and social change using a similar language. They believed in progress, but they didn't fall into the easy assumption that the private sector was solely responsible for it. And they believed they had new tools at their disposal, tools that would help them create a whole new wave of positive social change. Some of those tools involved technology, but many were low-tech in nature, ways of organizing resources or searching for solutions that didn't require a laptop.

The group did possess a slant toward technology in their collective background and training. Many of them had worked either directly or indirectly with the first wave of dot-com innovation during the 1990s. On some level, their high-tech roots were understandable. The one parish of modern life in which the religion of progress continues to reign supreme is, of course, that of digital technology. The crime rate can drop by 70 percent in less than a generation, but it's the launch of the iPad that gets on the cover of *Time*.

If you work around technology, it's a lot easier to be optimistic about future trends.

But those high-tech roots also proved to be a distraction. Skeptics assumed they were arguing that the Internet was a cure for everything. Some got typecast as "Net utopians," people for whom there was no social ill, no developing world tyrant, no medical crisis so intractable that it couldn't be toppled by throwing Facebook at it. No doubt some of the euphoria about the Internet's egalitarian promise was overstated, and some advocates did veer into genuine Net utopianism at times. But the people I was interested in were not evangelists for the Internet itself. For them, the Internet was not a cure-all; it was a role model. It wasn't the solution to the problem, but a way of thinking about the problem. One could use the Internet directly to improve people's lives, but also learn from the way the Internet had been organized, and apply those principles to help improve the way city governments worked, or school systems taught students. Sometimes computers happened to be involved in the process, but they weren't mandatory.

As I spent more time watching and thinking about this emerging movement, I began to realize that its political values did not readily map onto existing political categories. The people who most interested me were wary of centralized control, but they were not free-market libertarians.

They believed in the power of competition, but they also believed that some of society's most important achievements could not be incentivized with economic reward. They called themselves entrepreneurs but worked mostly in the public sector. They were equally suspicious of big government and big corporations. From a certain angle, they looked like a digital-age version of such nineteenth-century anarchists as Pierre-Joseph Proudhon or Mikhail Bakunin. Yet beneath that strange, hard-to-place eclecticism, a core philosophy seemed to unite all their efforts, even if it didn't fit comfortably into the existing categories of political belief.

From my own perspective, the most striking thing about these new activists and entrepreneurs was the personal chord that reverberated in me when I listened to them talk about their projects and collaborations—and their vision of the progress that would come from all that work. Here, at last, was a practical philosophy that resonated with the political riffs I'd been casually humming to myself over the past twenty years. In an age of great disillusionment with current institutions, here was a group that could inspire us, in part because they had attached themselves to a new kind of institution, more network than hierarchy—more like the Internet itself than the older models of Big Capital or Big Government.

This book is an attempt to take stock of this new vision,

Steven Johnson

to make the case that it is indeed a unified philosophy, and an original one. Most new movements start this way: hundreds or thousands of individuals and groups, working in different fields and different locations, start thinking about change using a common language, without necessarily recognizing those shared values. You just start following your own vector, propelled along by the people in your immediate vicinity. And then one day, you look up and realize that all those individual trajectories have turned into a wave.

I

THE PEER PROGRESSIVES

Do we believe in the individual,

or do we believe in the state?

—RAND PAUL

Few periods in history proved to be as turbulent as the forty years following the revolution of 1789 in France. Over just two generations, the country saw popular insurrection, regicides, new constitutions, a reign of terror, a world-conquering empire, and a temporary restoration of the old monarchy. Yet beneath all the political noise of that time, a steady rhythm persisted, a strange continuity from royals to radicals to republicans: the drumbeat for a world-class canal system. Despite the social turmoil, French engineers managed to complete a national network of canals by 1830, after decades of debate over the plan. It was an impressive achievement by any measure, but its completion happened to coincide with the emergence of the first successful industrial railroads in England and Germany. The canals of France were exquisite objects of design and engineering, but were obsolete on delivery.

Determined to avoid another national embarrassment, the new head of the Corps des Ponts et Chaussées (Corps of

Bridges and Roads), Victor Legrand, working with the brilliant engineer Claude-Louis Navier, set out to create a national railway system that would put the British and German railroads to shame. "What a fine role for the state," he announced, "if it can take charge and plan the main lines . . . and by this means of rapid, long-distance communications bring about the full integration of our fine country." The result of their collaboration would be perhaps the most iconic symbol of state planning ever built: an orderly, geometric series of rail lines radiating out across the nation from the center point of Paris. Where British and German rail lines followed chaotic, sinuous paths, defined by local topographies and the uncoordinated plans of competing rail companies, the French system was elegant and eminently legible, the industrial embodiment of what Pascal earlier had called *l'esprit géométrique*. It came to be known as the Legrand Star.

Navier's key engineering principles in the building of the Legrand Star all involved the reduction of local granularity in the service of geometric efficiency: Hills and curves would be avoided at all cost; grades were kept below five millimeters per meter. Winding rivers, valleys, hamlets, mountain ranges—all the local variations that humans had organically woven into their transportation routes, all the jumbled mess of the actual world—all this was to be ignored in the name of the straight line. In part, this was an aesthetic principle, but it was, on paper at least, also a matter of efficiency: Straight lines and level grades meant that people and goods could be transported across the country at higher speeds, with lower fuel costs. And in times of war, the same efficiencies would accelerate the nation's troop movements.

The *esprit géométrique* did not, in the end, live up to all of its promise. When the Franco-Prussian War erupted in 1870, Bismarck was able to transport nearly twice as many men to the front lines as the French could, despite the fact that the Prussian rail system was a hodgepodge of lines, run by more than fifty distinct private firms and public administrators. But the hodgepodge turned out to have a crucial advantage over the simplicity of the Legrand Star, what we would now call, in network theory, redundancy. The jumbled web of the German lines meant that troops could be conveyed to the war zone via six different lines, while the French military had to route the entire operation through a single line from Paris to Strasbourg. The heavily centralized nature of the Legrand Star also meant that reservists reporting from the provinces to the west and south had to travel through Paris before heading off for battle; Bismarck, on the other hand, had multiple rail centers at his disposal, which enabled more direct trips to the front lines, despite the decidedly ungeometric design of the railway systems.

Think of the Legrand Star as a kind of shorthand symbol for the ways that states like to organize the world. They concentrate power in a central location; they make the peripheries, the edges of the network, feeder systems for the main core; they simplify; they favor broad strokes over unpredictable swerves; they prefer master planners over local knowledge. They look best from above.

Twenty years ago, the Yale political science professor James C. Scott began investigating this way of interpreting and organizing the world. Trained as a scholar of Southeast Asian history, Scott published his first books on the semianarchic peasant societies of Malaysia and Vietnam, communities whose loose, flexible organizational structures were becoming increasingly anachronistic in a world dominated by massive states and corporations. These studies led Scott to investigate the mirror image of those small collectives—the mind of the state—to better understand what those peasant communities were up against. His research eventually took him through the invention of proper names, the development of grid-based systems of private property, the creation of majestic but sterile planned cities like Brasília, and the darker narratives of authoritarian terror executed in the name of utopian progress that dominated the first half of the twentieth century.

Scott ultimately published his survey as a book called *Seeing Like a State*. In it, the word that Scott returns to again and again is "legible." "The very concept of the modern

state," he writes at one point, "presupposes a vastly simplified and uniform property regime that is legible and hence
manipulable from the center." For the modern state to exist,
it must transform the vernacular and the idiosyncratic into
a set of standardized units that can be properly analyzed
and understood from a geographic and conceptual distance.
Without legibility, the state is blind. But making things legible means reducing their complexity, their nuance, their
crooked lines.

The legibility and centralization of classic state vision
have their benefits. Reducing individual humans to a data-
base of standardized statistical units helped create a revolution in public health that has lengthened and improved our
lives in countless ways. The French railway system may not
have defeated Bismarck, but it did allow the French to enjoy
the convenience of high-speed rail, in the form of the TGV,
decades before the rest of the world did. In many cases, as we
will see, the simplifications of the centralized state model
were necessary evils, given the technological limitations of
their time.

But by the middle of the twentieth century, the Legrand
Star model of state vision had come under attack from multiple opponents, occupying different points on the political
spectrum. Most famously, the Austrian economist Friedrich
Hayek demonstrated that centralized planning inevitably

confronted an information bottleneck, as the vast complex-
ities of a society were condensed down to a legible form that
a small coterie of planners could use as a basis for their
sweeping decisions. "The peculiar character of the problem
of a rational economic order," Hayek wrote, "is determined
precisely by the fact that the knowledge of the circumstances of which we must make use never exists in concentrated or integrated form, but solely as the dispersed bits of
incomplete and frequently contradictory knowledge which
all the separate individuals possess." Markets found a way
around this problem through the distributed signaling of
prices. If a shortage of (or new demand for) one particular
product arises—let's say it's the flash RAM so central to
mobile computing—the increase in price will trigger a wave
of altered behavior through the marketplace, with some
firms setting out to manufacture more flash RAM to meet
the demand, other firms reducing their dependence on it by
shifting to alternative storage systems, while other firms
increase the price of their mobile phones. But the individual
actors in that system can be utterly ignorant of the original
disruption in the supply of flash RAM. The price signaling
of markets, in Hayek's brilliant formulation, was a "system
of telecommunications," a way of solving the complex problem of constantly changing economic needs without reducing the whole mess down to a simplified central plan.

Hayek would go on to become a patron saint of the libertarian right, but his critique of Legrand Star planning had an unlikely ally in the American progressive urbanist Jane Jacobs, who followed Hayek's evisceration of Soviet planning with an equally devastating critique of master planners such as Robert Moses and the lifeless (and deadly) housing projects that had sprouted like concrete wildflowers in the postwar years. By replacing the local, intimate, improvisational balance of a city sidewalk with the bird's-eye view of automobile-centric planning, Moses and his peers were destroying the connective tissues of urban life. "Objects in cities—whether they are buildings, streets, parks, districts, landmarks, or anything else—can have radically differing effects, depending on the circumstances and contexts in which they exist," Jacobs wrote. "City processes in real life are too complex to be routine, too particularized for application as abstractions. They are always made up of interactions among unique combinations of particulars, and there is no substitute for knowing the particulars."

To a certain extent, these turned out to be blows from which the central planners never fully recovered. Today, in the United States at least, it is taken as a given that markets are the adaptive, innovative, responsive system in our culture, and that government and other public-sector institutions invariably must suffer from the Legrand Star syndrome. Even the most adamant libertarian still thinks we need some form of centralized government, of course. But the bureaucracies and Hayekian bottlenecks of the central planners are tolerated as a kind of necessary evil. Even the political Left works within the assumption that the private sector drives change and progress; the public sector, at best, creates safety nets.

Yet within a few months of Jacobs's launching her first volley against the titans of urban planning, a young researcher across the country was sketching a diagram that would ultimately find a way around Hayek's bottleneck.

In the mid-1950s, a Polish-born engineer named Paul Baran took a job at Hughes Aircraft while working on his graduate degree in engineering through night classes at UCLA. His work at Hughes gave him intimate access to the nascent technology of nuclear war—specifically the control systems that allowed the military to both detect inbound missiles and launch first strikes or retaliations. Years later, he would recall his horror at watching Hughes bid on the control system for the new Minuteman missile. "I was scared shitless," he later told Stewart Brand, in a *Wired* magazine interview, "because you had all these missiles that could go off by anyone's stupidity. The technology was never to be trusted."

As the cold war intensified after the launch of Sputnik in 1957, Baran took a new job at the RAND Corporation, where he got involved in a project to design a new command-and-control architecture for military communications. Baran was concerned that a nuclear detonation would disrupt high-frequency communications, so he began tinkering with a model whereby the military could hijack "ground wave" communications between broadcast stations, with each station relaying the message to others along the chain. The approach seemed more promising than the heavily centralized communications model employed by the then monopoly AT&T. The telephone system at the time was the communications equivalent of the Legrand Star: all the in-

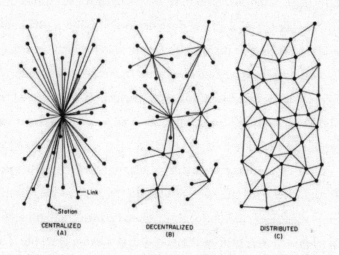

CENTRALIZED
(A)

DECENTRALIZED
(B)

DISTRIBUTED
(C)

Centralized, Decentralized, and Distributed Networks

telligence lay at the center of the network, which made it equally vulnerable to attack; take out the main switching centers of the phone network, and the entire system went offline. In 1960, Baran began sketching out a different, "fishnet" model for the system, one that didn't involve a central core but instead relied on a dense network of connections to shuffle information across the country. He compared three different models: a pure Legrand Star, an AT&T-style network with multiple cores, and a truly distributed net (see previous page).

The distributed model turned out to be far more resilient than the others. If one node went down, information could find its way to its ultimate destination through a different set of nodes, just as Bismarck's troops had used the fishnet-style railway lines of Germany in 1870 to get more men to the French border. In Baran's web, the center didn't need to hold. There was no center. That absence was what gave it its power and flexibility.

Baran's system included one additional conceptual breakthrough. Because the system was digital, not analog, individual messages could be easily broken up into smaller pieces and then reassembled at the end of the system with perfect fidelity. This, too, accentuated the resilience of Baran's architecture; even the messages themselves were mini-networks of data, with each partial message finding its own way across the broader network. Baran called his approach "distributed adaptive message block switching." A few years later, the Welsh computer scientist Donald Davies hit upon a similar scheme, independent of Baran. He anointed the message fragments with the slightly more Anglo name of "packets," and the general approach "packet switching." The metaphors stuck. Today, the vast majority of data circling around the globe comes in the form of message fragments that we still call packets. Years after both Baran and Davies had published their seminal papers, Davies jokingly said to Baran, "Well, you may have got there first, but I got the name."

In the late 1960s, packet switching became the foundation of ARPANET, the research network that laid the groundwork for the Internet. The ARPANET design relied on several radical principles that broke with existing computing paradigms. ARPANET was what we would now call a peer-to-peer network, as opposed to a client-server or mainframe-terminal network. Traditionally, networks had involved centralized mainframes that contained far more processing power and storage capacity than the less advanced terminals connected to them. The intelligence in the network, in other words, was centralized; decisions about what kind of information should be prioritized in the network were executed in these dominant machines. ARPA-NET, on the other hand, was a network of equals, of peers. No single machine had authority over the others. In this sense, the intelligence of the system was said to exist at the "edges of the network," not at its core.

Several years after the launch of ARPANET, Vint Cerf and Bob Kahn designed the TCP/IP protocols that became the common language of the Internet—the global network of networks. Their design included two crucial new principles that effectively increased the diversity of the network. First, TCP/IP was specifically engineered so that other networks could communicate through the Internet via "gateways." Those new networks could use their own proprietary languages if they wanted; as long as messages were encoded in TCP, the gateways would transmit the information, even if the computers used to transmit it didn't understand the contents of the message. Cerf and Kahn also designed the protocol so that new networking protocols could be developed on top of TCP/IP, so that future computer scientists could design new ways of communicating without having to engineer an entirely new global network from scratch. The two most common uses of the Internet—e-mail and Web browsing—relied on this crucial attribute, using distinct new protocols to exchange data on top of TCP/IP. We now think of these layers as "stacks" of different software platforms, arrayed on top of each other like so many layers at an archaeological site.

The creation of ARPANET and TCP/IP were milestones on many levels. They now rightly occupy a prominent place in the history of computing, communications, and globalization. But in a strange way, they should also be seen as milestones in the history of political philosophy. The ARPANET was a radically decentralized system that had somehow emerged out of a top-down government agency. The Baran Web showed that states did not have to create Legrand Stars, with their geometric lines and heavy centers. They could also create more fluid, dynamic structures that lacked hierarchies and centralized control. The ARPANET was like Hayek's marketplace, with "dispersed bits of information" and no central authority, but somehow, against all odds, it wasn't an actual marketplace. Hayek had metaphorically envisioned price signaling as "systems of telecommunications" that shared vital information throughout the network of buyers and sellers; here was an actual telecommunications system that did the same thing, except it managed to pull off the trick without prices.

One could be tempted to object at this point that states had a much longer history with decentralized control, in the form of democracies. It is true that the ability to choose your leaders through the electoral process pushes control to the edges of the system, compared with monarchies or other autocratic regimes. Yet almost all democracies, once the elections are over, are traditional centralized states. They are decentralized in the way they choose the people at the center of the Legrand Star; but once the votes are tallied, the authority and control and decision making come from the top. But the Internet, like Hayek's marketplace, is not a democracy. There are no popularly elected supercomputers that determine how the network is to be used. The users of the Internet did not vote for a small group of leaders to decide, for instance, what the search interface should be. We assumed that interesting search solutions would come from the edges of the network: from nonprofit organizations, and hackers, and academics, and start-ups.

In fact, the Internet is not even, strictly speaking, a "bottom-up" system, as it is conventionally described. Bottom-up implies a top, a leadership that gets its support from below, a kind of reverse hierarchy. But the network that Baran, Cerf, and others designed was a network of peers, not a hierarchy. No single agency controlled it absolutely; everyone controlled it partially.

Decentralization, peer-to-peer networks, gateways, platform stacks—the principles that Baran, Davies, Cerf, Kahn, and others hit upon together in the 1960s and 1970s provided a brilliant solution to the problem of sharing information on a planetary scale. Tellingly, the solution ultimately outperformed any rival approaches developed by the marketplace. Billions of dollars were spent by private companies trying to build global networks based on proprietary standards: AOL, CompuServe, Prodigy, Microsoft, Apple, and many others made epic efforts to become mainstream consumer networks in the late 1980s and early 1990s. They were all defeated by a set of networking standards—TCP/IP, the e-mail protocols of POP and SMTP, and the Web standards of HTML and HTTP—that were effectively public property: collectively developed and owned by no one, or by everyone. This was the stunning coda to Hayek's career: he won a Nobel Prize by explaining how markets shared information much more effectively than centralized states, but when it came time to build a global system for sharing information, the ultimate solution came from outside the marketplace.

It took many years for the brilliance and durability of that solution to become fully recognized. But here we are more than half a century after Paul Baran first sketched that fishnet diagram, and the principles that he helped discover continue to connect *Homo sapiens* on a scale never before seen in the history of the species. That epic success has, inevitably, led to new possibilities. Slowly but steadily, much like the creation of the Internet itself, a growing number of us have started to think that the core principles that governed the design of the Net could be applied to solve different kinds of problems—the problems that confront neighborhoods, artists, drug companies, parents, schools. You can see in all these efforts the emergence of a new political philosophy, as different from the state-centralized solutions of the old Left as it is from the libertarian market religion of the Right. The people behind these movements believe in government intervention without Legrand Stars, in Hayek-style distributed information without traditional marketplaces. Rand Paul's rallying cry was too simple; progress is not just a question of choosing between individuals and the state. Increasingly, we are choosing another path, one predicated on the power of networks. Not digital networks, necessarily, but instead the more general sense of the word: webs of human collaboration and exchange.

Yet as promising as it is, to date this underlying philosophy has been diffuse, not fully defined in the minds of its practitioners. For some time now, I have been casting around to try to find the right name for these emerging activists and thinkers, the umbrella word or phrase that captures their worldview. I toyed around with neologisms— "netarian" sounded too tech-based; "ambitarian," from the Latin for "edge," was too obscure. Ultimately I realized there was a word that had been staring me in the face all along: *peer*. It was a word that traveled nicely between high-tech and low-tech domains. The Internet and the Web (and almost all social networks that have been built on them) are peer-to-peer networks. The scholar Yochai Benkler had written eloquently of open-source projects such as Linux or Wikipedia as models of what he called "peer production." But *peer* in the social sense was an evocative and powerful word in its own right: your equals, the ones whose respect and judgment you seek.

So this is now what I call myself—and many of my friends and heroes—when asked about our political orientation. We believe in social progress, and we believe the most powerful tool to advance the cause of progress is the peer network. We are peer progressives.

In the early 1990s, two married staffers at Save the Children, Jerry and Monique Sternin, traveled to Vietnam at the behest of its government. They had been invited to tackle one of the most pressing health problems confronting the nation: at the time more than 50 percent of Vietnamese children suffered from chronic malnutrition. Because physiological development is so dependent on proper nourishment, childhood malnutrition is one of those problems that reverberates for a lifetime: brains and bodies deprived of the right proteins and vitamins during those crucial early years ultimately create adult brains and bodies that do not tap their full human potential.

The Vietnamese government had given the Sternins six months to show results in the field, so they set out quickly to a handful of rural villages, where they worked with Nguyen Thanh Hien, a local volunteer. But the Sternins did not descend on those communities with the usual imperious style of many foreign aid groups. They did not arrive as experts in nutrition, educating the locals on the best diets for their children, or strategies for ensuring that the water supply was safe. This was not a matter of style or multicultural sensitivity; it was a matter of practical effectiveness. The Sternins knew that "expert" strategies imported from other cultures had a very poor track record of long-term adoption. The aid workers would chopper into the village, give their lectures, wait for the heads to start nodding, then move on to the next community. And after a few months, the locals would revert to their old ways.

The Sternins decided to take another approach. They watched and listened, instead of lecturing. They assumed that the locals were the true experts in this context, that a sustainable solution to the problem of child malnutrition lay inside the community. It wasn't something that could be parachuted in from the outside, like a stash of energy bars. Their approach was inspired by a book that had just been published by a Tufts University professor, Marian Zeitlin, who had studied nutrition patterns in poor communities in the United States. Zeitlin had discovered that even the most destitute communities had successful outliers: families that had figured out a way to raise healthy children despite their bleak circumstances. Zeitlin called these individuals "positive deviants"—individuals or families who somehow managed to deviate from the health norms of the community in a good way.

In the Vietnamese villages, the Sternins first had to find the outliers, which was itself a tricky enough problem. They used traditional tools of measurement—weighing infants and tracking their growth, week by week—but they also relied on word of mouth, asking locals if they knew of families that seemed to produce unusually well-fed children. The Sternins quickly discovered that the villages had a meaningful collection of families that deviated positively from the nutrition norms. Then the Sternins' task turned to a different, more ethnographic, form of detective work: figuring out what exactly the outliers were doing. They studied their daily routines and compared them with the behaviors of families that had not broken free from the malnourishment cycle. As the months passed, the Sternins turned up a consistent set of behavioral patterns that were unique to the positive deviants. They fed their children via smaller, more frequent meals. ("There's only so much rice that a starving child's stomach can hold," Jerry Sternin used to say.) The positive deviants deviated from the taboos of social hierarchies, too: they pulled small shrimp and crabs from the rice paddies, and included them in their children's meals, despite the fact that shellfish were considered low-class in the culture. The shrimp and crab supplements might have turned up a few noses, but the cost of social prestige was more than made up for by the weight gain: they added critical proteins and vitamins to the children's diets.

Having isolated the essential behavioral patterns behind the positive deviance, the Sternins then set about to spread the word about the patterns, to amplify them in the community. Here again the Sternins relied on the existing village network. Instead of evangelizing the shrimp/crab diet through some kind of external authority, the Sternins encouraged the outliers to spread the word themselves. They also created subtle incentives that promoted the diet, holding daily sessions of medicinal-food training. Gatherings were indistinguishable from a fun, delectable repast with peers. As a price of admission, the villagers would have to bring shrimp, crabs, and sweet potatoes that they'd collected from the rice paddies; in return, they'd be treated to a satisfying meal.

The Sternins' earlier results were so impressive that they were allowed to continue the program for longer than their initial six-month term. After two years, they found that malnutrition had declined by somewhere between 65 and 85 percent in the villages. And the generation born after the Sternins had left remained at the same elevated levels of nutritional health, a sign that the success of the Sternins' project was not just a onetime intervention. The shrimp/crab diet had moved from the edges of the network to the mainstream. Once deviant, it had become the new norm.

The Sternins' engagement with those Vietnamese villagers was a decidedly low-tech affair. The most advanced piece of technology the locals ever directly interacted with was the scale used to weigh their children. But in a real sense, the Sternins were building a Baran Web in their search for positive deviants; instead of dictating laws from above, they worked within the existing peer networks of those communities in discovering, spreading, and rewarding the solutions that the villagers themselves had developed. The Sternins were not there to provide outside expertise; they were there to amplify the expertise that already existed in the community. The villagers had the tools they needed to feed their children more successfully; they just needed a little help sharing those tools through a wider network of their peers.

Twenty years after the Sternins arrived in Vietnam, the peer networks that they built while living in the rice paddies are now flourishing in communities all around the world. Some of these networks are deliberately choreographed by a group of activists or social entrepreneurs or government agencies; some of them are spontaneously forming in industries—like journalism—in which the larger institutions are failing and being replaced by interlinked networks of smaller, more nimble units. Seen as a whole, they make a compelling case for continued progress in our society. But to see them as a unified group, we need to step back and define their key attributes.

Peer networks involve several crucial elements, most of which involve concepts from network theory that have been refined over the past twenty years. They are decentralized in their control systems; no single individual or group is "in charge" of the system. The networks are dense, in that they involve a large number of participants, with many interconnections between them. They are diverse, in that the individual participants that make up the network bring different values or perspectives to the system. Peer networks emphasize open exchange over private property; new ideas are free to flow through the network as they are generated. And they incorporate some mechanism for assigning value to the information flowing through the network, promoting the positive deviants and discouraging the negative ones, creating incentives for participation in the network, or steering the system toward certain goals. (Crucially, that value need not take the form of traditional monetary units.) In addition, many of these networks come in layers or stacks, with new platforms of collaboration and exchange built on top of earlier platforms.

Seen together, the attributes happen to resemble closely (though not perfectly, as we will see) the design of the Internet itself, but they can be built into low-tech webs of collaboration as easily as digital platforms. Educational institutions, NGOs, local communities, protest movements—many different kinds of groups have created organizational structures or decision-making processes that rely on these principles.

This is one crucial way in which peer-progressive values are distinct from the stereotype of cyber-utopianism. There is nothing intrinsic to the peer-progressive worldview that says social problems can be wished away with some kind of magical Internet spell. For starters, many peer networks do not involve the Internet at all. (Think of the Sternins' search for positive deviance in the peer networks of rural Vietnam.) While the design of the Internet embodies most peer-progressive principles, its powers can be easily exploited by top-down, hierarchical organizations. The Internet makes it easier to build peer-based solutions to social problems, but it does not make them inevitable.

Peer networks are, in fact, much older than the Internet. The trading towns of the early Renaissance—Venice, Genoa, Istanbul—adhered to peer-network principles in much of their social organization: they lacked both big government and big corporations; they relied on densely populated urban streets where people from many cultures converged. And while they were trade centers driven by the exchange of private goods, the lack of mature intellectual property laws meant that new ideas and innovations were free to flow through the network.

The fact that those early trading towns are widely considered the birthplace of modern capitalism (in its pre-industrial form, at least) is no coincidence. True markets display almost all of the core principles of the peer-progressive worldview. As Hayek rightly observed, they work their magic through a decentralized network that can be wonderfully diverse in the cultures and perspectives it connects. Markets are also systems defined by the generative power of exchange, though over time that exchange has come to involve more restrictive concepts of intellectual property. When we talk about the importance of start-up culture in generating prosperity and innovation, we are in effect talking about the market's ability to generate new ideas at the edges of its network: in the garages and coworking spaces and grad school dropouts. All of which adds up to a simple but important truth: To be a peer progressive is to believe in the power of markets.

Up to this point, the peer progressives might as well be just another set of Hayek disciples—what are sometimes derisively called "the wired libertarians." But several crucial differences exist. For one, as Ned Beatty pointed out in a movie about a very different kind of network, twenty-first-century marketplaces are dominated by immense, hierarchically organized global corporations—the very antithesis of peer networks. The global marketplace that they have helped create is indeed a wonderful thing, but the power that has consolidated in the corner offices of those behemoths is not. When markets lead to vast financial discrepancies whereby the 1 percent controls more than 40 percent of the wealth in the society, that consolidation of economic power inevitably counteracts at least some of the decentralized structure of the marketplace. In other words, unlike many strands of older progressive movements, peer progressives genuinely like free markets; they're more ambivalent about CEOs and multinational corporations. When they look out at a globe with so much power concentrated in a handful of economic oligarchs, their reaction is like Gandhi's quip about Western civilization: "What do we think of free markets? We think they would be a good idea."

Unlike traditional libertarians, peer progressives do not believe that markets are capable of satisfying all of our human needs. This is where the experience of the Internet has been particularly instructive. When it came time to satisfy, on a global scale, that basic need for communication and the sharing of knowledge, the best solution turned out to emerge from open collaborative networks, not from private competition. The world is filled with countless other needs—for community, creativity, education, personal and environmental health—that traditional markets do a poor job of satisfying. Put another way, "market failures" are not just the twenty-year storms of major recessions or bank implosions. Markets are constantly failing all around us. The question is what you do when those failures happen. The pure libertarian response is to shrug and say, "That's life. A market failure will still be better in the long run than a big government fiasco." The traditional liberal response is to attack the problem with a top-down government intervention. The Right says, in effect, "Read your Hayek." The Left sets about to build a Legrand Star.

The peer-progressive response differs from both these approaches. Instead of turning a blind eye to market failures, it assumes that these problems are widespread, and actively seeks them out as the central focus of its agenda. Instead of building a large government agency to combat the problem, it tries to build a peer network around it, a system of dense, diverse, and decentralized exchange. Sometimes these interventions are supported by government funding; sometimes they are supported by charitable contributions; sometimes they involve Wikipedia-style contributions of free labor; sometimes they draw resources from private markets in creative new ways. In effect, they create Hayek-compatible solutions in the blank spots that the market has overlooked.

You could see one of those blank spots being filled if you happened to be a passenger riding on the Staten Island Ferry one bleak December morning in 2010. As the commuters sipped their coffee and flipped through the *Daily News* in one of the ferry's upper cabins, a twenty-one-year-old woman named Anne Marsen, dressed in brown corduroys and a loose turquoise blouse, suddenly launched into an exuberant dance routine, a weirdly captivating mix of calisthenics, break dancing, and pantomime baton tosses.

To the amused stares of the ferry regulars, Marsen twirled and pranced across the length of the boat, then into the streets of lower Manhattan, and ended up spinning between two empty park benches in front of City Hall.

This strange outburst of urban performance art was documented by Jacob Krupnick, a Brooklyn filmmaker who had met Marsen a year before, while shooting a promotional clip for a fashion show. Krupnick's video of Marsen's improvised performance was a test shoot for an ambitious but seemingly impractical idea for a music video that Krupnick had been toying with. The premise was impractical on multiple levels. For starters, Krupnick had no experience shooting music videos, and the music in question had been recorded by an artist that Krupnick had never met. Just as problematic, the musician happened to be the remix artist Girl Talk (aka Gregg Gillis), whose albums—composed entirely of elaborate samples from other musicians—had been released on the obscure label Illegal Art, a name that gives a reasonable sense of its relationship to existing intellectual property laws. (*The New York Times* once said that Girl Talk's music was a "copyright lawsuit waiting to happen.") Suffice it to say that there was no marketing budget at Illegal Art to fund a music video for Girl Talk.

But there was one additional complication. Krupnick's vision was not about making a standard pop song music video.

He wanted to make a single, continuous video to accompany the entire Girl Talk album. He wanted, in short, to make a seventy-one-minute-long video for an artist he didn't know and a label that had no promotional budget.

The history of creativity is replete with countless examples of talented and visionary individuals or groups who faced the same sort of impasse that confronted Jacob Krupnick. They had an idea for a new form of artistic expression that lacked an obvious commercial market, either because the ideas were too radical or because they relied on new formal conventions or genres that challenged existing modes, or because the creator was disconnected from mainstream circles of funding and support. (Krupnick, of course, hit the trifecta with his seventy-one-minute video.) Art that probes the boundaries of accepted ideas or taste rarely attracts enough of an audience to sustain itself financially. We have the phrase "starving artist" for a reason. And yet society as a whole benefits greatly from the network edges of experimental writing and music and theater and seventy-one-minute music videos. Subcultures expand the possibility space of our experience and our understanding; yesterday's underground is tomorrow's mainstream. (Recall that the whole practice of making music videos dates back, in part, to the experimental films Andy Warhol made with the Velvet Underground.) The fact that the marketplace reliably neglects creators like Jacob Krupnick is a kind of cultural market failure. Society is better off with dynamic, innovative cultures on their fringes, even if the short-term value of their work is not quantifiable. Yes, big-time record labels will occasionally dole out a six-figure advance to a new band on the basis of little more than YouTube buzz, and a visionary editor will buy a fragment of a novel from an unknown author. But most of those deals aren't struck until an artist has a proven track record, and building up that record takes time and money.

Perhaps the clearest evidence of the market failure around creative innovation lies in the sheer number and size of public-sector organizations devoted to encouraging the arts. We have the National Endowment for the Arts, the Doris Duke Charitable Foundation, and the MacArthur Foundation (and countless others) because many of us have come to the conclusion that the market on its own is not doing a sufficiently good job supporting people like Jacob Krupnick. We don't have a National Endowment for Detergents, because the private sector has proven to be perfectly adept at producing cheap and variegated quantities of household cleaning supplies. But the market has not performed as well with funding experimental or highbrow creative work. And so we build Legrand Stars to make up for that failure: large institutions with vast sums of money, endowed either by mega-rich patrons or the federal government, controlled by small committees of experts who decide which artists are worth supporting.

Historically, Jacob Krupnick would have been forced to choose among three paths: He could reconfigure his artistic vision to make it more compatible with the existing marketplace. He could seek out the patronage of a large institution or, in rare cases, a wealthy individual. Or he could carve out spare time to work on his project after-hours, while keeping a day job. Those were his options: go mainstream, find a wealthy benefactor, or turn his creative vision into a part-time hobby.

But Krupnick was facing this dilemma in 2010, which meant that he had another option: a website named Kickstarter. Founded in April 2009 by Perry Chen, Charles Adler, and Yancey Strickler, Kickstarter is perhaps the most successful of a new generation of "crowdfunding" sites that organize financial support for creative or charitable causes through distributed networks of small donors. On Kickstarter, artists upload short descriptions of their projects: a book of poetry that's only half completed, a song cycle that has yet to be recorded, a script for a short film that needs a crew to get produced. Kickstarter's founders defined "creative" quite broadly: technological creativity is welcome, as are innovations in such fields as food or design. More than a few microbreweries have been launched on Kickstarter, and one of the most successful early projects (in terms of funds raised) involved turning the iPod Nano into a wristwatch. When Jacob Krupnick first announced his project—which he had come to call *Girl Walk // All Day*—on Kickstarter, he uploaded his footage of Anne Marsen dancing on the Staten Island Ferry, along with a more detailed description of his ultimate goal:

> The idea behind *Girl Walk // All Day* emerged from our desire to expand the boundaries around the idea of the traditional music video, which usually spans the length of a single track. This album-length piece will feature a talented group of dancers across a range of public and private spaces around New York City, turning the city's sidewalks and obstacles into part of an evolving improvisational dance routine. The piece will be available for free online in short, serialized segments and we also plan to screen the full-length film in public spaces, and at festivals, concerts, parties, and beyond, inspiring an interactive viewing experience that will evolve into a series of dance parties around the globe.

Once creators have defined their project on Kickstarter, they then have to establish two crucial items: how much money they need to complete their project, and what reward they are offering their patrons in return for their financial support. The financial goals are crucial to Kickstarter's success: when would-be patrons pledge money to support a project, their credit card isn't charged until the creators receive enough pledges to reach their goal. In other words, there is no risk of contributing funds to a project that doesn't end up happening because the creators ultimately failed to raise enough money. In fact, almost half of the projects on Kickstarter fail to reach their goals. If you happen to back one of those projects, your pledge is erased, as if you'd never contributed in the first place.

The reward system is also crucial to Kickstarter's success, and is itself a space of great creativity on the site; the only real limitation on the rewards is that they not be strictly monetary in nature—in other words, you can't promise your patrons a financial return on their investment. So instead of a musician offering her supporters a cut of future sales, she offers free tickets to the launch party or a private performance in their living room or a signed copy of a CD. Borrowing a page from most nonprofit pledge drives, Kickstarter makes it simple to create tiered rewards for patrons according to how much money they contribute. In Krupnick's case, his rewards ranged from a thank-you on the project website for contributions up to $15, all the way to a private dance lesson from Anne Marsen (along with an invitation to the wrap party and other perks) for contributions over $500.

Krupnick's project went live on Kickstarter on January 31, 2011. He asked for $4,800 to support six days of shooting, the subsequent editing and production of DVDs, and the projectors for public screenings. Within two months, the project was fully funded. By the time Krupnick stopped taking contributions, he had raised $24,817. Nearly six hundred individual backers had given money to the project. Almost half of those contributions were less than $30.

In late 2011, Krupnick completed the seventy-one-minute version of *Girl Walk // All Day*. He had begun the year with an improbable idea, no funding, and no connections. His only asset was an inspiring video of an unknown twenty-one-year-old dancer who herself had no professional experience. By the end of the year, that trailer video had landed Anne Marsen a role on the hit TV series *The Good Wife*. And in early 2012, when *Spin* magazine announced its list of the most innovative music videos of the year, *Girl Walk // All Day* was at the very top of the list, beating out videos by Beyoncé, Adele, and the Beastie Boys.

How novel is the Kickstarter crowdfunding approach? If you look at it exclusively in terms of the donations themselves, it doesn't seem to be a genuine paradigm shift. Any regular listener to NPR will tell you that public media has long relied on small donations induced with tiered rewards. Starting with the Howard Dean campaign of 2004, politicians have become adept at generating small contributions from vast numbers of supporters using real-time goals and social networks. But if you tried to diagram these systems, they would look more like the Legrand Star than the Baran Web: all those donations flowing into the central mother ship of NPR or the Dean headquarters. The money comes from the edges of the network, but the center still decides what programs should air, or what the campaign strategies should be.

Kickstarter, on the other hand, does away with the center altogether. Both the ideas and the funding come from the edges of the network; the service itself just supplies the software that makes those connections possible. The donors decide which projects deserve support. There are no experts, no leaders, no bureaucrats—only peers. New creative ideas don't need to win over an elite group of powerful individuals huddled in a conference room, and they don't need to win over a mass audience. All they need is an informal cluster of supporters, each contributing a relatively small amount of money. They can build that network of support directly through Kickstarter, but they can also augment it with their own social connections. (Solicitations for

Kickstarter projects flow regularly through the networks of Facebook and Twitter.) There have always been far more people on the planet willing and capable of spending fifty dollars to support an artist than there were people willing to spend a million. It was just too hard to round up all those small donors, so artists who couldn't find support in the private market (or who didn't want to play by those rules for political reasons) focused their fund-raising pitches on moguls and princes. But as Clay Shirky has powerfully argued over the years, the Internet is brilliant at reducing the organizational costs of creating and maintaining groups, particularly casual groups that are defined by loose affiliations.

In Jacob Krupnick's case, he was trying to put together a group of people who were so interested in seeing a seventy-one-minute music video that they'd be willing to front the money to pay for it—with zero possibility of additional return if the video somehow became a breakout success. He needed, in other words, seed investors who were cool with the fact that they had no hope of seeing an economic flower. An unlikely group, to be sure—yet somehow Krupnick assembled six hundred of them in a matter of months.

To a traditional economist, there's something baffling about the lack of an "upside" in the Kickstarter donation. By strict utilitarian standards, the vast majority of Kickstarter donors are wildly overpaying for the product. No music video—however long, however large the typeface they use to thank donors personally in the credits—is worth a hundred dollars, particularly a music video by an unknown director that hasn't, technically, been made yet. So why does the contribution get made? The return on the Kickstarter investment can't be measured by the conventional yardstick of utilitarian economic theory. People contribute for more subtle, but just as powerful, reasons: the psychological reward of knowing that their money is helping cultivate another human's talents; the social reward of being seen in public doing just that. Drawing on the work of Lewis Hyde, the writer Kevin Kelly calls this kind of activity the Web's "gift economy." Kelly described this phenomenon in a *Wall Street Journal* op-ed published shortly after the dot-com bubble burst in 2000:

> As the Internet continues to expand in volume and diversity without interruption, only a relatively small percent of its total mass will be money-making. The rest will be created and maintained out of passion, enthusiasm, a sense of civic obligation, or simply on the faith that it may later provide some economic use. High-profile portal sites like Yahoo and AOL will continue to consolidate and demand our attention (and maybe make some money), while millions of smaller sites and hundreds of millions of users do the heavy work of creating content that is used and linked. These will be paid entirely in the gift economy.

Wikipedia, to use only the most obvious example, is powered entirely by the gift economy, mostly in the form of the free labor contributed by its authors and editors. Unlike Wikipedia, though, Kickstarter is a hybrid system. Designed to support creative work that the market does not value, the service is itself a for-profit company, funded by some of the leading venture capitalists in the world. It is a private firm whose shareholders will likely see a very tidy return on their investment. And yet the firm creates a platform where individuals can share resources without the need for financial return on their seed funding. The company itself is modern capitalism at its most advanced: a venture-backed Web start-up. But what it creates is a gift economy.

There is something undeniably baffling about this, almost as if some nineteenth-century industrial giant had created a factory that produced socialist collectives. I suspect that if you were to describe Kickstarter to most people a few years ago, the idea would have seemed like a pipe dream at best: Imagine a company that asks ordinary people to give money to unknown artists to support work that they haven't completed yet. And imagine that company makes a comfortable profit for its founders and investors. Most people would dismiss the idea as a fantasy—a lovely idea in theory that would never work in practice. At best, it might draw a small fringe audience and support a handful of artists. But supporting creative work on a large scale, they'd argue, requires big institutions: either media conglomerates or giant philanthropic institutions. That's just the way it works.

But as Jacob Krupnick discovered, Kickstarter is no pipe dream. Two and a half years after Chen, Adler, and Strickler launched the site, they announced that Kickstarter was on track to raise roughly $200 million for artists in a single fiscal year. The entire annual budget for the National Endowment for the Arts is $154 million. The question with Kickstarter, given its growth rate, is not whether it could ever rival the NEA in its support of the creative arts. The new question is whether it will grow to be ten times the size of the NEA.

The peer-network approach to funding worthy causes is hardly limited to Kickstarter alone. Over the past few years, dozens of services have emerged, targeting different problems with similar crowdfunding techniques. Some, such as the Australian service Pozible, share Kickstarter's focus on creative ideas; others tap into large networks of patrons to support nonprofit charities, such as the site Causes.com, cofounded by Sean Parker, of Napster and Facebook fame. Some focus on specific fields, as in the education crowdfunding service DonorsChoose, which connects individual teachers to potential patrons who want to support innovative projects in the classroom. In 2011, roughly $1.5 billion passed through crowdfunding sites. And that's not including the political campaigns that have been revolutionized by crowdfunding techniques over the past decade.

Kickstarter and its fellow crowdfunding sites work because the services are built on peer-progressive values. Kickstarter, for instance, took on an existing problem that markets had traditionally fumbled—how do we find and support interesting new creative forms—and radically increased both the density and diversity of the participants. It gave thousands of creative people direct access to the wallets of millions of potential patrons. Before Kickstarter, if you were the sort of person who was interested in supporting fledgling artists, it was actually quite hard to meet fledgling artists, and almost impossible to meet them in bulk. Kickstarter changed all that. It increased the density of links connecting artists and their would-be supporters, and it increased the diversity of those groups.

Selection, too, is crucial to the genius of Kickstarter, thanks in large part to the patron's simple act of supporting one project over many potential rivals. Interesting, provocative, polished, ambitious ideas get funding; boring or trivial or spammy ones don't. At last count, 48 percent of Kickstarter projects do not reach their funding goal, and thus raise zero dollars. This is, as they say, a feature, not a bug.

All of which sounds like market mechanisms in precisely the mode that Hayek described: the consumers collectively weeding out the bad ideas through the magic of paying for things. As Hayek would have put it, the knowledge of which artists deserve support "never exists in concentrated or integrated form, but solely as the dispersed bits of incomplete and frequently contradictory knowledge which all the separate individuals possess." But for the precise problem that Kickstarter is trying to solve, the fact that it relies on the gift economy makes it more efficient than a traditional market. Patrons are not basing the amount of their contributions on how much something is worth to them in terms of short-term practical use. Nor are they betting that the idea they are backing will eventually reach a mass audience. Were they traditional consumers or investors, the selection pressure would be toward much more predictable fare. But the gift economy turns all that on its head. It promotes risk, experimentation, cultural slow hunches that may not turn into commercial ideas for twenty years.

Because Kickstarter the company belongs to the private sector, because it has shareholders who are motivated by traditional market incentives, it may be tempting at this point to assume that Kickstarter itself is just a roundabout argument for libertarian values. Sure, you may need gift economies to support fringe or early-stage creative ventures, but eventually the market will come up with a for-profit company that will create a space where those gift exchanges can happen. In other words, the marketplace is so relentlessly innovative that it will eventually come up with a non-market solution if that's what's required.

Yet there is nothing in Kickstarter's DNA that says it has to be a for-profit company. We could easily decide as a society that the $200 million Kickstarter is disbursing is not nearly enough to support the kind of creative innovation we need in our culture. At which point, the government could create its own Kickstarter and promote it via its own channels, or it could use taxpayer dollars as matching grants to amplify the effect of each Kickstarter donation. This, in a nutshell, is the difference between a libertarian and a peer-progressive approach. The libertarian looks at Kickstarter and says, "Great, now we can do away with the NEA." The peer progressive says, "Now we can make the NEA look more like Kickstarter."

What is ultimately important about Kickstarter is not whether it is a for-profit company or a creature of the gift economy or some interesting hybrid—what is important is the social architecture of the service. A social architecture is a set of rules and conventions that govern the way a group interacts: official laws or casual customs, geographic and conceptual arrangements. Some social architectures—like those of most corporations or religious organizations—are profoundly hierarchical; others—such as communes or trading towns—have more fluid, horizontal structures. Some create very strict limitations on the flow of information through the group. (Think of the National Security Agency, or the Coca-Cola formula.) Others are predicated on the open flow of information, as in most university cultures. You could say that the "native" social architecture of the online world is the peer network. The Internet and the Web were built, and are maintained, by peer networks: dense, diverse, and distributed networks of open collaboration and exchange. And while it is certainly possible to use the Internet to strengthen your hierarchical organization, the Internet seems to have a bias toward peer networks, if only because it makes it so much easier to assemble them. We didn't have Kickstarter or Wikipedia before the Web came along because the organizational costs of connecting all those people were prohibitive.

The fact that the Net is biased toward peer-network architectures is one critical reason why many current examples of peer progressivism have digital technology at their core. The new ideas took root there more easily. But now they are ready to disseminate.

To date, the most prominent examples of network architectures influencing real-world change have been the decentralized protest movements that have emerged over the past few years: MoveOn, Arab Spring, the Spanish Revolution, Occupy Wall Street. These movements have been fascinating to watch, and like their smaller equivalents on Facebook (the "Like" campaigns for various causes), they have succeeded brilliantly at expressing a popular dissatisfaction with the status quo, building awareness for a particular injustice, and on occasion raising money. But they have all proved to be somewhat disappointing at actually proposing new solutions and making those solutions reality. They are brilliant at swarming, building feedback loops of energy and attention. They are less adept at steering. The grand spectacles of Occupy or Arab Spring have turned out to be something of a distraction, averting our eyes from the more concrete and practical successes of peer networks. Kickstarter, for instance, is not a platform for expressing outrage at the woeful state of arts funding. It is a platform for getting things done.

Could it be that peer networks won't perform as well outside the grounds of Web 2.0 technology? Systems based on pure information are clearly more amenable to the experimentation, decentralization, and diversity of peer networks than more material realms are. It's harder to get those kinds of groups to gather in a barn or a city hall than it is to assemble them virtually. Moving bits around is far easier than doing the—sometimes literal—heavy lifting of civic life: building reservoirs or highways or jet engines that don't explode when birds fly into them. A skeptic might plausibly say, Sure, if you want to conduct an online poll, or crowdsource a city slogan, the Internet is a great leap forward. But if you want to do the real work, you need the older tools.

But every material advance in human history—from the Great Wall to the Hoover Dam to the polio vaccine to the iPad—was ultimately the by-product of information transfer and decision making. This is how progress happens: some problem or unmet need is identified, imaginative new solutions are proposed, and eventually society decides to implement one (or more) of those solutions. The circulation of ideas and decisions in that cycle is ultimately as important as the physical matter that is transformed in implementing the solutions themselves. Yes, it was crucial to the passengers on US Airways flight 1549 that the plane's engines had been forged via the staggering physical energy and immense financial expense of a jet engine production cycle. But it was just as important that someone, somewhere, had decided that it would be a good idea to make sure those engines could survive bird impacts. The information that made that decision possible, and the social architecture that allowed its wisdom to spread—all these were every bit as important to the safe landing on the Hudson as the physical act of building the engines. Yes, peer networks can't do everything. They're just a better way to decide. But you can't have progress without good decisions.

To be a peer progressive, then, is to believe that the key to continued progress lies in building peer networks in as many regions of modern life as possible: in education, health care, city neighborhoods, private corporations, and government agencies. When a need arises in society that goes unmet, our first impulse should be to build a peer network to solve that problem. Some of those networks will rely heavily on digital network technology, as Kickstarter does; others will be built using older tools of community and communication, including that timeless platform of humans gathering in the same room and talking to one another.

The fact that the peer network does not fit easily into the traditional political categories of the Left and Right should not be mistaken for some kind of squishy, "third way" centrism. It is not the moderate's attempt to use Big Government and Big Labor to counterbalance the excesses of Big Corporations. Living strictly by peer-progressive values means rethinking the fundamental structures of some of the most revered institutions of modern life; it means going back to the drawing board to think about how private companies and democracies are structured. It is, as the political writer Micah Sifry likes to say, not a matter of finding a middle ground between Left and Right, but rather finding a way forward. This is why it is so important that these principles not be confused with simple Internet utopianism. There's nothing radical in taking an existing institution and putting it on the Internet; that's a job for an IT department, not a political upheaval. What peer progressives want to see is fundamental change in the social architecture of those institutions, not just a Web strategy.

There is a utopian strain to this vision, to be sure—utopian in the sense of both unchecked optimism and a certain lack of real-world practicality. Rebuilding the social architecture of the U.S. electoral system, to give one obvious example, may not be an achievable goal in the short or even medium term. But Wilde had it right: "A map of the world that does not include utopia is not even worth consulting." It may well not be possible to implement peer-progressive values on the scale some of us would like—in our lifetimes, at least. But that doesn't mean we shouldn't try to imagine what those larger changes would look like, if only to give us something epic and inspiring as a lodestar, guiding our more incremental movements.

Still, the peer network is not some rarefied theory, dreamed up on a commune somewhere, or in a grad school seminar on radical thought. It is a practical, living, evolving reality, one that is already transforming dozens of different sectors. It underlies the dominant communications system of our time, along with some of the most significant social movements. This is why it is such an interesting and encouraging time to build on these values. We have a theory of peer networks. We have the practice of building them. And we have results. We know that peer networks can work in the real world. The task now is to discover how far they can take us.

II

PEER NETWORKS AT WORK

The Maple Syrup Event

New Yorkers are no strangers to strong odors, but several years ago, a new aroma began wafting through the city streets at unpredictable moments, a smell that was more unnerving than the usual offenders (trash, sweat, urine) precisely because it was so delightful: the unmistakable scent of maple syrup. The scent would saturate an entire neighborhood, like some kind of sweet miasma.

The maple syrup smell was fickle, though. Every month or two, Manhattanites in various neighborhoods on the west side of the island would encounter it for a few hours before it faded. The maple syrup smell would drape itself over Morningside Heights one afternoon, disappear for a few weeks, then reemerge in Chelsea for a few passing hours before vanishing again. Fearing a chemical warfare attack, presumably from the Aunt Jemima wing of Al Qaeda, thousands of New Yorkers reported the smell to the authorities. *The New York Times* first wrote about it in October 2005; local blogs covered each outbreak, augmented by firsthand accounts in their comment threads.

The city quickly determined that the smell was harmless, but the mystery of its origins persisted for three years. In an earlier era, getting the word out about the innocuous nature of the smell would have been a challenge for the authorities. It was hardly an emergency, after all, and the smell itself affected only a small number of New Yorkers at unpredictable times. They might have reached out through the papers and news radio to spread the word, but even those stories would be fleeting.

But this was 2005, and the city had an extraordinary resource at its disposal, one that had been created only a few years before: the 311 service, where New Yorkers can do everything from reporting potholes to inquiring about alternate-side-of-the-street parking rules to complaining about unusual smells, all by dialing three digits on their phones. Launched in March 2003, 311 now fields on average more than 50,000 calls a day, offering callers information about more than 3,000 services: school closings, recycling rules, homeless shelters, park events, pothole repairs. The service has translators on call to support 180 different languages. Since its launch, 311 operators in New York have fielded more than 100 million calls.

Having those 311 call centers in place offered the city a direct medium with which to communicate with citizens worried about the strange breakfast smells saturating their neighborhoods. During "maple syrup events," as city offi-cials came to call them, 311 operators were instructed to ex-plain to callers that the smell was harmless, and that they should go about their business as usual.

But the true genius of 311 lies in the fact that it is a two-way system: 311 callers get the information they need from the service, but the city also learns from the information the callers contribute. Each call is tagged and categorized and mapped so that the city can detect patterns in the otherwise overwhelming chaos of a teeming metropolis: a cluster of noise complaints surrounding a late-night bar; a growing litany of requests for more children's events in a neighborhood; potholes, broken swings, abandoned cars— all the molecular needs and obstacles and opportunities that ultimately determine the overall health of the urban superorganism.

This capacity for pattern detection turned out to be cru-cial to the strange case of the "maple syrup events." Several years after these odd smells began wafting over the city, a few members of the 311 team had an idea: Those maple syrup calls into the 311 line, they realized, weren't simply queries from an edgy populace. They were clues. The data the city collected weren't just making them aware of an ol-factory puzzle. It was also the key to solving that puzzle.

On January 29, 2009, another maple syrup event commenced in northern Manhattan. The first reports triggered a new protocol that routed all complaints to the Office of Emergency Management and the New York City Department of Environmental Protection, which took precise location data from each syrup smeller. Within hours, inspectors were taking air-quality samples in the affected regions. The reports were tagged by location and mapped against previous complaints. A working group gathered atmospheric data from past syrup events: temperature, humidity, wind direction, velocity.

Seen all together, the data formed a giant arrow aiming at a group of industrial plants in northeastern New Jersey. A quick bit of shoe-leather detective work led the authorities to a flavor compound manufacturer named Frutarom, which had been processing fenugreek seeds on January 29. Fenugreek is a versatile spice used in many cuisines around the world, but in American supermarkets it is most commonly found in the products on one shelf—the one where they sell cheap maple syrup substitutes.

Fourteen months after the maple syrup mystery was solved, Mayor Michael Bloomberg paid a visit to the 311 call center, which is housed in a city building in the warrens of downtown Manhattan, just a few blocks east of Ground

Zero. With its high ceilings, playful Flor tiles underfoot, and dual LCD monitors on every desk, the main call center room feels a lot like a Web start-up, until one registers the steady murmur of the fifty "customer service professionals" working the phones. Mounted on one wall is an oversize dashboard, with chunky blue, red, and green LED pixels tallying the day's inflows by city department: calls waiting, maximum waiting time, agents on call—and the most important statistic of all, service level, which reports the percentage of calls that are answered within thirty seconds. Bloomberg's visit commemorated 311's hundred millionth call, and for the photo op, the mayor fielded one call himself. As it happened, the caller recognized Bloomberg's voice; amazingly, he turned out to be an old colleague from the mayor's investment banking days at Salomon Brothers. Even the biggest cities have small towns buried within them.

There was something fitting in this unlikely connection, since 311 is designed to re-create some of the human touch of small-town culture in the context of a vast metropolis. Callers are guaranteed to reach a live person within thirty seconds, after a brief prerecorded message summing up the day's parking regulations (a major topic of 311 queries) and other relevant news. Crucial as well to the 311 ethos is the idea of civic accountability: by giving New Yorkers an easy venue to report broken streetlights or graffiti or after-hours construction, the service helps them solve the problems they see in their own neighborhoods.

As useful as 311 is to ordinary New Yorkers, the most intriguing thing about the service is the data it supplies back to the city. Each call is logged, tagged, mapped, and entered into the Customer Relationship Management system to make it available for subsequent analysis. The CRM records everything from street addresses to the city agency involved to the nature of the inquiry or complaint (everything from library hours to syrup-smell panic attacks, and so on). In some cases, 311 simply lets the city collect and share information more efficiently about needs that were obvious to begin with. On snow days, for instance, call volume spikes precipitously, which 311 anticipates by adding prerecorded information about school closings and parking rules to the opening message. (On February 25, 2010, during one of the worst snowstorms in New York's history, 311 fielded more than 269,000 calls.) The ethnic diversity of the city means that religious holidays trigger reliable surges in call volume, with questions about government closings and parking regulations.

But the service also allows the city to detect patterns that had otherwise gone unnoticed. Some of those patterns are macro-trends. After the first survey of 311 complaints

ranked excessive noise as the number one source of irrita-
tion among residents, the Bloomberg administration insti-
tuted a series of noise-abatement programs, going after the
offenders whom callers complained about most often. Simi-
larly, clusters of complaints about public drinking in certain
neighborhoods have led to crackdowns on illegal social
clubs. Some of the discoveries have been subtle but brilliant. For example, officials now know that the first warm
day of spring will bring a surge in use of the city's chloro-
fluorocarbon recycling programs. The connection is logical
once you think about it: The hot weather inspires people to
upgrade their air conditioners, and they don't want to just
leave the old, Freon-filled units out on the street.

The 311 hive mind is deft not just at detecting reliable
patterns but also at providing insights when the normal
patterns have been disrupted. Clusters of calls about food-
borne illness or sanitary problems from the same restau-
rant now trigger a rapid response from the city's health
department. And during emergencies, callers help provide
real-time insight into what's really happening. After US Airways flight 1549 crash-landed on the Hudson, a few callers
dialed 311 asking what they should do with hand luggage
they'd retrieved from the river. The city had extensive plans
for its response to an urban plane crash, but dealing with
floating luggage was not one of them. Within minutes they

had established a procedure for New Yorkers who wanted to turn in debris they'd recovered from the river. This is the beauty of 311. It thrives on the quotidian, the predictable: the school-closing queries and pothole complaints. But it also plays well with black swans.

If anyone still wonders whether the 311 concept is here to stay, New York's hundred millionth call should dispel all doubts. So, for that matter, should the other hundred-plus 311s now in operation across the United States. For millions of Americans, it's a three-digit dial that has become almost as second-nature as its neighbors on the keypad, 411 and 911. Perhaps even more exciting is the new ecosystem of start-ups—inspired in part by New York's success and empowered by twenty-first-century technology—that has sprouted up to imagine even more innovative ways for residents to report local needs or issues. A service called SeeClickFix lets users report open fire hydrants, dangerous intersections, threatening tree limbs, and the like. (A similar service, FixMyStreet, launched in the UK several years ago.) In proper Web 2.0 fashion, all reports are visible to the community, and other members can vote to endorse the complaints. Another start-up, BlockChalk, has released an iPhone app that uses GPS data to let users create public notes tagged to specific locations. CitySourced, an angel-backed start-up, has partnered with the city of San Jose to serve as a high-tech front end for its 311 system. A New York–based site called UncivilServants collects reports and photos of government workers abusing parking rules around the city and ranks the top offenders by department. (The worst abuser, by a wide margin, is the NYPD.) There's even a new Australian urban reporting site called, memorably, It's Buggered Mate.

Taken together, all this adaptive, flexible urban reporting points the way toward a larger, and potentially revolutionary, development: the crowdsourced metropolis, the city of quants.

Systems such as 311 and its ilk are the peer-progressive response to the problem that all great cities invariably confront: the problem of figuring out where the problems are. In the language of *Seeing Like a State*, 311 makes the city legible from below. It doesn't set its priorities from above; it doesn't get bound up in official definitions or categories. It is designed to listen to the word on the street and develop its priorities and responses from what it learns. This is why the story of the maple syrup events is so instructive: 311 was not designed to detect and track neighborhood-wide outbreaks of sweet breakfast smells. That particular skill was not part of the original specs for the service. But 311 turned out to be brilliant at it, precisely because it was designed as a peer network, in which new ideas and patterns could be generated at the edges, without any official decree from the center.

It should be said that 311 is not a purely decentralized system. There are both literal and figurative headquarters, where the call center is located. In this sense, it is a hybrid form, somewhere between the pure peer network and the older state model. The 311 service vastly increases the number of participants in the system, and gives them the opportunity to set priorities for the city's interventions. But those interventions are still triggered via a top-down mechanism.

To a certain extent, that top-down element may be inevitable. Perhaps someday a descendant of 311 will allow small groups of citizens to self-organize teams to repair potholes on neighborhood streets, bypassing direct government intervention altogether. But even if citizen street repair never comes into being, there are countless ways that true peer-to-peer civic interaction will flourish in tomorrow's neighborhoods. Consider the elemental needs of transportation. A pedestrian standing at any intersection in Manhattan has at least four modes of transportation to choose from: cab, bus, subway, or foot. (The list could potentially be longer: hitching a ride in a private car or taking a pedicab, for instance.) In some cases, there are dozens of bus and subway lines within a few blocks, and hundreds of taxis. Each is a potential data point—the F train that's twelve minutes behind schedule, the six cabs just around the corner looking for fares. This, too, is an information problem that can be solved through a peer-network approach. Make every vehicle and (willing) individual a node in the broader network, and let the network learn the fastest and most efficient routes, just as the Internet itself routes packets around the globe. Some of the data might come from official city sources, but much of it would be distributed in nature. When New York's Taxi and Limousine Commission installed television screens and credit card machines in all taxis, they also installed GPS devices that communicate vast amounts of information back to the TLC. Every second of every day, 13,000 cabs send real-time data on location, travel speeds, and whether they have customers or not. Combine that data with live information from mass transit, individual data uploaded from mobile phones, and even Yelp-style reviews of the most interesting streets for window shopping—and the decision of how to get from point X to point Y becomes far more interesting.

In other words, 311 is just the beginning: As technologies evolve, all this pooling and sharing and analysis of data will allow cities to become increasingly sophisticated in solving urban problems. These solutions won't be about just the edges of the network communicating local needs back to the center, but the edges sharing resources and tools with one another.

A few years ago, old friends (and neighbors) of mine decided to embark on what was originally going to be a small renovation of their basement. Somehow, over the course of the construction work, the foundation of their nineteenth-century brownstone was compromised—to the point where one structural engineer advised that they move out of the house until the problem had been fixed. There were a few days of panic, but after the contractor did some work shoring up the basement, my friends were told that the house had been temporarily stabilized—assuming that there were no abnormal vibrations or earthquakes in the vicinity. (In Brooklyn, earthquakes were not exactly in the forecast.) And in another month or two, when the final work was done, the house would actually be more stable than it had originally been.

But then, just a few days after the structural engineer advised them that it was safe to return to their home, my friend learned while chatting with a neighbor that the city was days away from starting a major sewage pipe replacement project on their very block, closing down the street, with jackhammers and diggers carving out a ten-foot hole in the pavement. In other words, a veritable symphony of abnormal vibrations thundering through their house for six straight weeks.

There's an important lesson in this story—and not just about the importance of hiring the right contractor. The real lesson lies in the news value of that information about the city's construction plans for the street. For my friends, that little nugget of information was arguably the single most vital headline they could possibly have read that week, far more important than anything going on in Iraq or in the U.S. campaign season, much less in Britney Spears's custody battle. It was news that had significant financial and safety implications for their entire family. Yet despite its urgency, the news had arrived on their doorstep via the word-of-mouth network of two neighbors gossiping together.

All of this would seem to reinforce an observation Chris Anderson (of *Wired* and The Long Tail fame) made several years ago, part of what he called the "Vanishing Point theory of news":

Our interest in a subject is in inverse proportion to its distance (geographic, emotional or otherwise) from us. For instance, the news that my daughter got

> a scraped knee on the playground today means more
> to me than a car bombing in Kandahar.

Anderson's framework points to something crucial in understanding how local communities work. The conventional assumptions about priority and relevance get turned on their head when we are talking about people's lived environments: their front stoops and playgrounds and commuting routes. In the case of my friend's compromised foundation, the crux of the problem was one of information management: my friend needed the crucial information about an upcoming sewer repair, and the city needed the crucial information about a house that was likely to collapse once the jackhammers started up. On the face of it, the news of imminent sewer repair was trivial, but in a vanishing-point context, it was very nearly life or death.

Legrand Stars tend to founder when confronted with these vanishing-point problems, because taking on the bird's-eye view of the state means erasing or normalizing—or simply ignoring—all those local variations. It's hard enough for the city government to figure out which streets need sewer repair, much less figure out which of those streets happened to contain houses whose foundations have been structurally compromised. Making the city legible has historically meant eliminating those micro-local details, which meant that city governments often failed to meet the shifting and unpredictable needs of its citizens. But peer networks don't have to erase those details. The relationship between the government and its citizens stops being a mass relationship—a command-and-control headquarters towering over a faceless aggregate—and becomes a two-way conversation. In a world with 311-style networks shaping the flow of information through the community, it becomes far easier for the citizens to learn about upcoming sewer repair on their block—and for the government to learn which sewer projects should be postponed.

When these peer networks work, they challenge most of our easy assumptions about government bureaucracies. We deplore bureaucracies for understandable reasons: they are slow, inflexible, wedded to arbitrary conventions, and incapable of responding to individual needs. They are natural outgrowths of the Legrand Star, in that the needs of the general population have to be condensed down into legible packages so that the small number of individuals within the state can manage and make sense of the city's complexity. But government programs, like 311, that draw on peer-network structures don't suffer from the same constraints. Done well, they can even outperform the market. Several years after the launch of 311, New York City commissioned a third-party firm to compare the customer satisfaction of

311 users with that of users of other call centers in both the public and private sectors. The 311 customer satisfaction ratings came out on top, barely edging out hotel and retail experiences but beating other government call centers, such as the IRS's, by a mile. (At the very bottom of the list, not surprisingly, cable companies.) Extracting information from, or reporting problems to, a government agency is traditionally a recipe for psychic pain, because bureaucracies are so incompetent at addressing individual needs. Yet 311 has managed to break from that ignoble tradition—precisely because it is a network, not a bureaucracy.

The 311 system was designed and is maintained by public-sector employees, but the peer networks that will come to shape our communities will not necessarily emerge exclusively from governments. Already, some promising hybrid models have appeared: SeeClickFix has begun offering free dashboards for local governments, with a premium service available for a monthly fee. The service also bundles together its user-generated reports and e-mails them to the appropriate authorities in each market. It's an intriguing hybrid model, in which the private sector creates the interfaces for managing and mapping urban issues while the public sector continues its traditional role of resolving those issues.

Another interesting approach can be found in Open311,

a new project spearheaded by the OpenPlans organization. Open311 is, in a sense, the Linux of 311 systems: a standardized, open-source platform that cities and private entities can use to track "issues with public space and public services." The first iterations of the Open311 system are now live in San Francisco and in Washington, D.C., where users can report basic quality-of-life complaints: the usual suspects of potholes, garbage, vandalism. Unlike New York's current 311 system, Open311 is designed to be a true "read-write" platform: anyone can use the system to contribute new data, and anyone can extract data as well. That means outside parties can develop new interfaces, both reporting problems and visualizing those that have already been reported. In the current 311 paradigm, each new city is the equivalent of a different operating system, because the structures of the underlying data vary so radically from place to place. But with Open311, an app built for San Francisco can be instantly ported over to work with Washington, D.C.'s data. It becomes, in effect, a meta-peer network: the software that creates a more distributed and diverse community of civic participation is itself built by a more distributed and diverse community.

Now imagine two extra layers: a layer that allows members of the community to propose solutions to the problems or new opportunities that arise via Open311. And then a layer that allows individuals to fund those projects directly through small donations. In other words, imagine a system that discovers the problems with 311 and solves them with Kickstarter. This is what a group of social designers in Helsinki are beginning to do, under the all-too-perfect code name of Brickstarter. Here's how it might work in practice: Say, for the sake of argument, that there's an empty lot in a city neighborhood that's overrun with weeds and beer bottles. Locals complain about the lot via some form of 311: it's an eyesore; kids are breaking into it and smoking cigarettes; the barbed-wire fence is a potential hazard. Those complaints accumulate and eventually become a meaningful cluster on the map of the neighborhood, a red flag that says, "There's a problem at this address." At that point, the Brickstarter network invites proposals from neighbors and other relevant parties: one group of green thumbs suggests a community garden; a local contractor offers to build a playground; others suggest paving it over and turning it into a parking lot. The projects set their funding goals and rewards, and the wider neighborhood community selects a winner by pledging money to support it. It would be peer networks all the way up the chain. A peer network builds tools that let a peer network of neighbors identify problems or unmet needs in a community, while other networks propose and fund solutions to them.

Why isn't that a model that could productively supplement, if not partially replace, the existing democratic forms that govern most communities, in the developed world at least? Some might object that such a system would favor wealthier communities over poorer ones: the empty lots would get converted into gleaming playgrounds in the neighborhoods with the time and money to contribute to Brickstarter; but they'd remain eyesores in the neighborhoods that were living hand to mouth. But the money you're donating to Brickstarter doesn't have to be spare change you have lying around in the bank account. Imagine, instead, that the money comes in the form of actual tax dollars. Instead of writing, say, a $10,000 check to the local tax collector, you'd be allowed to "spend" some portion of your taxes on the Brickstarter projects that you felt were worthwhile. Poorer neighborhoods would get tax credits that could be pledged toward Brickstarter projects, with the money itself coming from tax dollars collected in wealthier communities.

This is not a "thousand points of light"–style downsizing of government itself, to be replaced by purely elective charitable giving. The government would be just as big in terms of the tax dollars flowing through it. In fact, if Kickstarter's success is any indication, people might well be willing to pay higher local taxes if they felt they had a direct say in how they were being spent. But the government would be much less hierarchical in the decision-making process that governed the spending of all that money. The task of identifying and solving community problems would be pushed out to the edges of the network, away from the central planners. In the case of neighborhoods, of course, that movement to the edges makes perfect sense, because in real-world communities, the true experts are more often than not the people who actually live in those communities. Neighbors understand intuitively what's working and what's broken on their sidewalks and in their cul-de-sacs and parks. They know where the positive deviants live.

The Pothole Paradox

The problems and opportunities we find in local governance correlate very closely with the problems of journalism. Both domains share a defining quality: they involve problems of information management. How do you get the right information to the right people at the right time? How do you prioritize one piece of news over another? Chris Anderson's vanishing-point theory was originally conceived as an observation about the relative value of different kinds of news: the micro-local news of a daughter's scraped knee was more important to him than the story about a car bombing around the world—even though the latter story was, in absolute terms, clearly more significant.

When you think about local news, though, it quickly becomes clear that the vanishing-point theory is only part of the problem. It is true that, all other things being equal, news events that are nearer to us matter more than news events far away, which is why minor happenings such as a sewer repair can trump the results of a political election, given the right circumstances. But there is a corollary to this vanishing-point perspective, one that I have come to call "the pothole paradox." The logic goes as follows:

1. Say you've got a particularly nasty pothole on your street that you've been scraping the undercarriage of your car against for a year. When the town or city finally decides to fix the pothole, that event is genuine news in your world. But it is precisely the kind of news that traditional media has utterly failed to convey to its audience. You'll never open your local paper or tune in to the eleven-o'clock news and get word about a pothole repair on your block.

2. News about a pothole repair just five blocks from your street is the least interesting thing you could possibly imagine.

To use Anderson's vanishing-point framework, my kid scraping his knee at the playground might well be news in my household (though given my boys' habits at the playground, it would be pretty repetitive news), but news about Anderson's kid scraping her knee is as boring to me as, well, pothole repair five blocks away from my house.

The pothole paradox plays out with any number of different topics. The delicious Indian place that at long last opens up in your neighborhood; the notoriously dull science teacher who finally retires at the local public school; the come-from-behind victory staged by the middle school lacrosse team—all of these are potentially exciting events if they happen in the communities you inhabit, but they are mind-numbingly dull if they're one county over, much less on another coast.

The other complication here is that the correct scale of local news varies, depending on the nature of the news itself. Pothole repair may die out beyond a few blocks, but many happenings—crimes or political rallies or controversial real estate developments—reverberate more widely. Going local sometimes requires that you zoom in all the way to the block level, even all the way to the individual address. But sometimes you need to zoom out as well. Think of each news story as having a geographic zone of relevance: some stories stop at the end of the block; some remain relevant throughout the neighborhood.

The pothole paradox partially explains why the traditional journalistic institutions have such a poor track record of satisfying the demand for local news. No one wants to read a newspaper story about other people's potholes, and no one wants to tune in to a sports broadcast that covers a middle school football game from someone else's school district. A newspaper filled with other people's hyperlocal news is not a paper anyone wants to read. And even if there were an audience for it, a citywide paper filled with hyperlocal news is a practical impossibility. Local news fails the Hayek test, because it is made up of small bits of dispersed information that cannot be easily condensed and made legible by a centralized office of official experts. Even in their heyday, newspapers didn't have the staff to put a reporter on every elementary school in the city, or every construction crew. Big-city newspapers or television news anchors are the central planners of local news: a small group of people who have to reduce or ignore the local granularity of potholes and school principals and lacrosse games to produce something that will be legible to the widest possible audience.

But those limitations don't apply with a peer network. You aren't likely to learn about an upcoming sewer repair on your block if your news source is a centralized city paper. But as my friend discovered, you might well learn about it if your news source is another neighbor one block over whose street has been closed for the past week. The pothole paradox is a challenge when the news is coming via a Legrand Star. But if it's coming via a Baran Web, the rules are different. Not every individual bit of news information in the system has to be interesting to the entire audience, and the number of potential contributors to the system is large enough to put a potential reporter on every street corner. In a world where media are made by peers and not just papers, new kinds of journalism become possible.

The future of journalism has, of course, been the subject of great debate over the past half decade or so. The simplest way to understand what has happened over that period is this: the overarching system of news is transitioning from a Legrand Star to a Baran Web, from a small set of hierarchical organizations to a distributed network of smaller and more diverse entities. Because this transition involves the failure or downsizing of many of those older organizations, and because those organizations have, for the past few centuries at least, been our primary conduits of reported news and commentary, many thoughtful observers have seen that transition as a crisis and a potential threat. But from a peer-progressive point of view, the emerging news system looks like an improvement over the old order. It looks like opportunity, not crisis. But to understand why, we have to look at the old system with fresh eyes.

If you happened to be hanging out in front of the old College Hill Bookstore in Providence, Rhode Island, in 1987, on the third week of every month you would have spotted me walking into the bookstore several times a day. I wish I could tell you that I was making those compulsive return visits out of a passionate love for books. While I did in fact buy plenty of books during my college years, I was making those tactical strikes on the College Hill Bookstore for another reason. I was looking for the latest issue of *Macworld*.

I had learned from experience that new issues of the monthly magazine devoted to all things Macintosh arrived at College Hill reliably during the third week of the month. Yes, you could subscribe, but subscription copies tended to arrive a few days later than the copies in the College Hill Bookstore. So when that time of the month rolled around, I'd organize my week around regular check-ins at College Hill to see if a shipment of *Macworld*s had landed on their magazine rack.

This was obsessive behavior, I admit, but not entirely irrational. It was the result of a kind of imbalance—not a chemical imbalance, an information imbalance. To understand the peer-progressive take on the future of journalism, it's essential that we travel back to my holding pattern outside the College Hill Bookstore—which continued unabated, by the way, for three years. If we're going to have a responsible conversation about the future of news, we need to start by talking about the past. We need to be reminded of what life was like before the Web.

I made my monthly pilgrimages to College Hill because I was interested in the Mac, which was, it should be said, a niche interest in 1987, though not that much of a niche. Apple was one of the world's largest creators of personal computers, and by far the most innovative. But if you wanted to find out news about the Mac—new machines from Apple, the latest word on the upcoming System 7 or HyperCard, or any new releases from the thousands of software developers or peripheral manufacturers—if you wanted to keep up with any of this, there was just about one channel available to you, as a college student in Providence, Rhode Island. You read *Macworld*.

Even then, even if you staked out the College Hill Bookstore, waiting for issues hot off the press, you were still getting the news a month or two late, given the long lead times of a print magazine back then. Yes, if Apple had a major product announcement, or fired Steve Jobs, it would make it into *The New York Times* or *The Wall Street Journal* the next day. And you could occasionally steal a few nuggets of news by hanging around the university computer store. But that was basically the extent of the channels available to you.

When I left college and came to New York in the early nineties, the technology channels began to widen ever so slightly. At some point in that period, I joined CompuServe, and discovered that *MacWEEK* magazine was uploading its articles every Friday night at around six, which quickly became a kind of nerd version of appointment television for me. The information lag went from months to days. In 1993, *Wired* magazine launched, and suddenly I had access not only to an amazing monthly repository of technology news, but also to a new kind of in-depth analysis that had never appeared in the pages of *Macworld*.

Within a few years, the Web arrived, and soon after, I was reading a site called MacInTouch, which featured daily updates and commentary on everything from new printer driver releases to the future of the Mac clone business. Tech critics such as Scott Rosenberg and Andrew Leonard at Salon wrote tens of thousands of words on the latest developments at Apple. (I wrote a few thousand myself at the online magazine I had founded, FEED.) Sometime around then, Apple launched its first official website; now I could get breaking news about the company directly from the source, the second they announced it.

If my nineteen-year-old self could time-travel to the present day, he would no doubt be amazed by all the Apple technology—the iPhones and MacBook Airs—but I think he would be just as amazed by the sheer volume and diversity of the information about Apple available now. In the old days, it might have taken months for details from a John Sculley keynote to make it to the College Hill Bookstore; now the lag is seconds, with dozens of people live-blogging every passing phrase from a Steve Jobs or Tim Cook speech. There are 8,000-word dissections of each new release of OS X at the technological site Ars Technica, written with attention to detail and technical sophistication that far exceed anything a traditional newspaper would ever attempt. Writers such as John Gruber and Donald Norman regularly post intricate critiques of user-interface issues. The traditional newspapers have improved their coverage as well: think of David Pogue's reviews in *The New York Times*, or Dow-Jones's extended technology site, AllThingsD. And that's not even mentioning the rumor blogs.

Of course, *Macworld* is still around as a print magazine, but it also now has a website. On an average day, it publishes more than twenty different articles on Apple-related topics.

The story of technology news over the past twenty years is the story of a peer network coming into being, a system far more decentralized and diverse than anything that came before it. Unlike Kickstarter or 311, the peer network of tech news was not a single platform created by a few visionaries and entrepreneurs. The network evolved through thousands of small contributions, some of them building out the technical infrastructure of blogs and wikis and tweets, some of them contributing the actual reporting and commentary. The network of tech news mirrored the digital network it was covering. The results of all that labor are undeniable: by almost every important standard, the state of Mac news has vastly improved since 1987. There is more volume, diversity, timeliness, and depth.

When you hear people sound alarms about the future of news, they often gravitate to two key endangered species: war reporters and investigative journalists. Will the bloggers get out of their pajamas and head up the Baghdad bureau? Will they do the kind of relentless shoe-leather detective work that made Woodward and Bernstein household names? These are genuinely important questions, and I think we have good reason to be optimistic about their answers. But you can't see the reasons for that optimism by looking at the current state of investigative journalism in the blogosphere, because the peer network of investigative journalism is in its infancy. There are dozens of interesting projects being spearheaded by very smart people, some of them nonprofits, some for profit. But they are seedlings.

I think it's much more instructive to anticipate the future of investigative journalism by looking at the past of technology journalism. When ecologists go into the field to research natural ecosystems, they seek out the old-growth forests, the places where nature has had the longest amount of time to evolve and diversify and interconnect. They don't study the Brazilian rain forest by looking at a field that was clear-cut two years ago.

That's why the ecosystem of technology news is so crucial. It is the old-growth forest of the Web. It is the subgenre of news that has had the longest time to evolve. The Web doesn't have some kind of intrinsic aptitude for covering technology better than other fields. It just has an intrinsic tendency to cover technology *first*, because the first people that used the Web were far more interested in technology than they were in, say, school board meetings or the NFL. But that has changed, and is continuing to change. The transformation from the desert of *Macworld* to the rich diversity of today's tech coverage is happening in all areas of news. Like William Gibson's future, it's just not evenly distributed yet.

Consider another—slightly less nerdy—case study: politics. The first presidential election that I followed in an obsessive way was the 1992 race between Clinton and then President Bush. I was as compulsive a news junkie about that campaign as I was about the Mac in college: every day the *Times* would have a handful of stories about the campaign stops or debates or latest polls. Every night I would dutifully tune in to *Crossfire* to hear what the punditocracy had to say about the day's events. I read *Newsweek* and *Time* and *The New Republic*, and scoured *The New Yorker* for its occasional political pieces. When the debates aired, I'd watch religiously and stay up late, soaking in the commentary from the assembled experts.

That was hardly a desert, to be sure. But compare it with the information channels that were available to me after the 2008 election. Everything I relied on in 1992 was still around, of course—except for the late, lamented *Crossfire*—but it was now part of a vast new forest of news, data, opinion, satire—and perhaps most important, direct experience. Sites such as Talking Points Memo and Politico did extensive direct reporting. Daily Kos provided in-depth surveys and field reports on state races that the *Times* would never have had the ink to cover. Individual bloggers such as Andrew Sullivan responded to each twist in the news cycle; The Huffington Post culled the most provocative opinion pieces from the rest of the blogosphere. The statistician Nate Silver at the website Five Thirty Eight did meta-analysis of polling that exceeded anything Bill Schneider dreamed of doing on CNN in 1992. When the banking crisis erupted in September 2008, I followed economist bloggers such as Brad DeLong to get their expert take on the candidates' responses to the crisis. I watched the debates with a thousand virtual friends live-tweeting alongside me on the couch. All this was filtered and remixed through the extraordinary political satire of Jon Stewart and Stephen Colbert, whom I watched via viral clips on the Web as much as I watched them on TV.

What's more, the ecosystem of political news also included information coming directly from the candidates. Think of Obama's widely acclaimed Philadelphia speech responding to the controversy over his connection to the Reverend Jeremiah Wright, arguably one of the two or three most important events in the whole campaign. Eight million people watched it on YouTube alone. What would have happened to that speech had it been delivered in 1992? Would any of the networks have aired it in its entirety? Certainly not. It would have been reduced to a minute-long sound bite on the evening news. CNN probably would have aired it live, which might have meant that 500,000 people caught it, and of course Fox News and MSNBC didn't exist yet. A few serious newspapers might have reprinted it in its entirety, which might have added another million to the audience. Online, someone would have uploaded a transcript to CompuServe or The Well, but that's about the most we could have hoped for.

There is no question in my mind that the political news ecosystem of 2008 was far superior to that of 1992: I had more information about the state of the race, the tactics of both campaigns, the issues they were wrestling with, the mind of the electorate in different regions of the country. And I had more immediate access to the candidates themselves: their speeches and unscripted exchanges; their body language and position papers.

I think the political Web covering the 2008 campaign was so rich for precisely the same reasons that the technology Web is so rich: because it's old-growth media. The first wave of blogs were tech-focused, and then, for whatever reason, they turned to politics next. And so the peer networks of political coverage have had a decade to mature into their current state. What's happened with technology and politics is happening elsewhere, too, just on a different timetable. Sports; business; reviews of movies, books, restaurants—all the staples of the old newspaper format are proliferating online. There are more perspectives; more depth and more surface now. And that's the new growth. It's only started to mature.

In fact, I think in the long run, we're going to look back at many facets of old media and realize how much vital news we were missing in the old regime. Local news may be the best example of this. When people talk about the civic damage a community suffers by losing its newspaper, one of the key things that people point to is the loss of local news coverage. But I suspect in ten years, when we look back at traditional local coverage, it will look much more like *Macworld*, circa 1987. When I lived in New York City, I adored the City section of the *Times*, but every Sunday when I would pick it up, there were only three or four stories in the whole section that I found interesting or relevant to my life—out of probably twenty stories total. Yet every week in my neighborhood there would be easily twenty stories that I would be interested in reading: a mugging three blocks from my house; a new deli opening; a house sale; the baseball team at my kid's school winning a big game. *The New York Times* can't cover those things in a print paper not because of some journalistic failing on their part, but rather because the economics are all wrong: there are only a few thousand people potentially interested in those news events, in a city of 8 million people. There are metro-area stories that matter to everyone in a city: mayoral races, school cuts, big snowstorms. But most of what we care about in our local experience lives in the vanishing point. We've never thought of it as a failing of the newspaper that its metro section didn't report on a deli closing, because it wasn't even conceivable that a big centralized paper could cover an event with such a small radius of interest. But peer networks can. They can find a way around the pothole paradox.

So this is what the old-growth forests tell us: there is going to be more content, not less; more information, more analysis, more precision, a wider range of niches covered. You can see the process happening already in most of the major sections of the paper: tech, politics, finance, sports. It is possible that somehow investigative or international reporting won't thrive on its own in this new ecosystem, that we'll look back in ten years and realize that, somehow, almost everything in journalism improved except for those two areas. But there is substantial evidence to suggest that the venerable tradition of the muckraking journalist will be alive and well ten years from now: partly supported by newspapers and magazines, partly by nonprofit foundations and innovative programs like California Watch or ProPublica, and partly by enterprising bloggers who make a name for themselves by breaking important stories.

Skeptics who are less sanguine about the future of peer-network journalism have two primary concerns, one financial and one ideological. For more than a century, serious journalism has been financially supported by the massive profits newspapers accumulated, thanks in large part to the near monopoly they had on local advertising. That cash cow has disappeared, and as a result the old news organizations have closed their Baghdad bureaus, sliced their print publications in half, and shifted focus toward lighter service journalism. If serious reporting is going to prosper in this new era, who is going to foot the bill?

It is certainly true that the old business model of newspapers has run its course. It's almost impossible to maintain a monopoly in a peer-network world. But you can't judge the overall system simply by looking at the financial health of the old organizations. Peer networks may make it harder for large organizations to capture outsize profits, but they also vastly increase the efficiency of the overall information system. They support the news not by allowing a handful of big organizations to print money. Instead, they support the news by making it cheap.

Ecologists talk about the "productivity" of an ecosystem, which is a measure of how effectively the ecosystem converts the energy and nutrients coming into the system into biological growth. A productive ecosystem, such as a rain forest, sustains more life per unit of energy than an unproductive ecosystem, such as a desert. We need a comparable yardstick for information systems, a measure of a system's ability to extract value from a given unit of information. We can't fully appreciate the power of peer networks without a model of information productivity. The density and open exchange of a peer network drive up the information productivity of the overall system, because new ideas circulate quickly through the network and can be built upon or expanded at little cost. You can do more with less.

The overall increase in information productivity may be the single most important fact about the Web's growth over the past fifteen years. Think about it this way: Let's say it's 1995, and you are cultivating a page of "hot links," as we used to call them, to interesting discoveries on the Web. You find an article about a journalism lecture at Columbia University and you write up a quick description with a link to the Columbia website that promotes the talk. The information value you have created is useful exclusively to two groups: people interested in journalism who happen to visit your page, and the people maintaining the Columbia page, who benefit from the increased traffic.

Fast-forward to the present: You arrive at the lecture and check in at Foursquare, and tweet a link to a description of the talk. Set aside the fact that it is now much easier to make those updates via your smartphone, compared with the cumbersome process of updating your website circa 1995. What happens to the information you send out? It's the same number of characters, with the same message: I'm going to this lecture tonight. But the ultimate trajectory of that information is radically more complex than it would have been fifteen years before. For starters, it goes out to friends of yours, and into your Twitter feed, and into Google's index. The geo-data embedded in the link alerts local businesses who can offer your promotions through Foursquare; the link to the talk helps Google build its index of the Web, which then attracts advertisers interested in your location or the topic of journalism itself. Because that tiny little snippet of information flows through a more dense and diverse network, by checking in at the lecture you are helping your friends figure out what to do tonight; you're helping Columbia promote its event; you're helping a nearby bar attract more customers; you're helping Google organize the Web; you're helping people searching for information about journalism; you're helping journalism schools advertising on Google to attract new students. When text is free to flow and recombine, new forms of value are created, and the overall productivity of the system increases.

The new species emerging in the ecosystem understand this intuitively in ways that the older life-forms can't. ProPublica, the nonprofit news org that won a Pulitzer Prize in 2010 for its reporting, runs an Abbie Hoffmanesque line on its website entreating readers to "steal our stories." It sounds like it should be said with a wink, but ProPublica means it with a straight face. ProPublica has licensed its content under Creative Commons, so that whoever wants to publish its articles may do so, as long as this is done with a credit (and link to) ProPublica and all links that appeared in the original story are included. Instead of putting journalism under glass, the organization is effectively saying to its information, Go forth and multiply. The organization itself is designed explicitly on peer-network principles, but it also wants to make sure the information it produces flows through as wide a net as possible.

One of the reasons ProPublica can do this, of course, is that it is a nonprofit whose mission is to be influential, and not to make money. It seems to me that this is one area that has been underanalyzed in the vast, sprawling conversation about the future of journalism over the past year or so. A number of commentators have discussed the role of non-profits in filling the hole created by the decline of print newspapers. But they have underestimated the information productivity of organizations that are incentivized to connect, not protect, their words. A single piece of information designed to flow through the entire ecosystem of news will create more value than a piece of information sealed up in a glass box. You may well make more money if you build walls of copyright law around your information—though there is some interesting debate about even that—but there is no question you are diminishing the influence of your ideas when you erect those barricades.

And ProPublica, of course, is just the tip of the iceberg. There are thousands of organizations—some of them focused on journalism, some of them government-based, some of them new creatures indigenous to the Web—that create information that can be freely recombined into neighborhood blog posts or Pulitzer Prize–winning investigative journalism. ProPublica journalists today can get the idea for an investigation from a document on

WikiLeaks, get background information from Wikipedia, or download government statistics or transcripts from data .gov or the Sunlight Foundation. They can even get funding directly from Kickstarter. Once again, the power of the system comes not just from the individual peer networks, but from the way the different networks layer on top of one another.

You cannot measure the health of journalism simply by looking at the number of editors and reporters on the payroll of newspapers. There are undoubtedly going to be fewer of them. But there is every reason to believe that loss is going to be more than offset by the tremendous increase in information productivity we get from peer-networked news.

Peer-produced journalism may well be able to solve the funding problem without relying on advertising monopolies, but that still leaves the ideological question about the future of news. Some critics believe that a world in which news is divided up into much smaller and more diverse sources will inevitably lead to political echo chambers, where like-minded partisans reinforce their beliefs by filtering out dissenting views. Perhaps the most articulate and influential advocate for this belief is the legal scholar and Obama administration official Cass Sunstein, who describes the echo-chamber effect this way:

> If Republicans are talking only with Republicans, if Democrats are talking primarily with Democrats, if members of the religious right speak mostly to each other, and if radical feminists talk largely to radical feminists, there is a potential for the development of different forms of extremism, and for profound mutual misunderstandings with individuals outside the group.

When groups can filter their news by ideological persuasion, the long-term tendency is toward increased polarization and decreased consensus. Individuals' interpretations of the world get amplified and not challenged; the common ground of social agreement shrinks. When groups are exposed to a more diverse range of perspectives, when their values are forced to confront different viewpoints, they are likely to approach the world in a more nuanced way, and avoid falling prey to crude extremism.

Sunstein and his fellow echo-chamber skeptics are absolutely right to point to the social value of building links between diverse groups. A peer progressive would go even one step further. Diversity does not just expand the common ground of consensus. It also increases the larger group's ability to solve problems. The pioneer in this line of research is the University of Michigan professor Scott E. Page. He has spent the past twenty years building a convincing case for what he calls the "Diversity trumps ability" theory, demonstrating the phenomenon in sociological studies and mathematical models. Take two groups of individuals and assign to each one some kind of problem to solve. One group has a higher average IQ than the other, and is more homogeneous in its composition. One group, say, is all doctors with IQs above 130; the second group doesn't perform as well on the IQ tests, but includes a wide range of professions. What Page found, paradoxically, was that the diverse group was ultimately smarter than the smart group. The individuals in the high-IQ group might have scored better individually on intelligence tests, but when it came to solving problems as a group, diversity matters more than individual brainpower.

The problem-solving capacity that comes from diverse networks is one of the cornerstones of the peer-progressive worldview. In this sense, it deviates somewhat from older progressive traditions, in which diversity is championed in the name of social equality and tolerance. Peer progressives don't simply want diversity in governments and corporations and educational institutions because certain groups have historically been disenfranchised and thus deserve their proper representations in those organizations. (Though that certainly is a worthy cause.) We want diversity for another reason as well: because we are smarter as a society—more innovative and flexible in our thinking—when diverse perspectives collaborate. For the peer progressive, too, the emphasis on diversity does not revolve exclusively around the multicultural diversity of race or gender; it's as much about professional, economic, and intellectual diversity as it is about identity politics.

Diversity, then, is an undeniable virtue on multiple levels. The question is whether a peer-produced news environment creates more or less of it.

If you look at the overall system of journalism today, it seems preposterous to argue that there has been a decrease in the diversity of news and opinion, compared with the media landscape of the pre-Internet era. Every niche perspective—from the extremes of neo-Nazi hate groups to their polar opposites on the far Left—now has a publishing platform, and a global audience, that far exceeds anything they could have achieved in the age of mass media.

The echo-chamber critics would no doubt accept this description of the overall system. But they would counter that this systemic diversity leads, paradoxically, to a narrowing of individual perspectives, because people can now custom-tailor their news to a much smaller slice of the ideological spectrum. In the old days, the limited options meant that people were dependent on mass-media institutions—newspapers, the nightly news—that featured an array of topics and perspectives, thus serving up a more diverse information diet. To continue the culinary metaphor, contemporary peer-produced journalism is like one of those diner menus that go on for ten pages, which means that everyone can find their favorite dish and never experiment. Old media, on the other hand, is like a chef's tasting menu: far less choice, but because an expert chef is making the decisions for you, you end up expanding your palate by eating food you would have never ordered on your own.

This debate is ultimately about the relative power of filters versus links. In a world of peer-network media, you can use filters to narrow your perspective down to a vanishing point of interest. But that world is also densely interconnected, thanks to the peer-networked, hypertextual architecture of the Web and the Internet. For the echo-chamber argument to hold water, you have to believe that the connective power of the Web is weaker than the filters. Even the most partisan blogs are usually only one click away from their political opposites, whereas in the old world of print magazines or face-to-face groups, the opportunity to stumble across an opposing point of view was much more unusual.

Fortunately, we no longer have to speculate about these things. Over the past few years, a number of rigorous sociological experiments have shown that our concerns about echo chambers online are greatly exaggerated. In 2010, two University of Chicago business professors, Matthew Gentzkow and Jesse M. Shapiro, published an exhaustive study of the echo-chamber effect in various forms of media, as well as in real-world communities such as neighborhoods or voluntary organizations. Their study created what they called an "isolation index"—a measure of how much the average reader or browser or citizen in a given environment isolates himself from differing political perspectives. If the average newspaper reader jumped from Tea Party op-eds to socialist commentary in a newspaper diet, then newspapers as a medium would warrant a very low isolation rating. If the average right-wing Web surfer locked himself up on the Fox News site and never left, then the Web would score much higher on the isolation index.

Gentzkow and Shapiro found that the Web produced an isolation index that was right in the middle of the spectrum of recent media platforms. The echo-chamber effect online was slightly stronger than it was among readers of local newspapers or cable news channels. And it was slightly lower than the isolation index for national newspapers. Strikingly, Web users who had extreme political viewpoints were much more likely than moderate users to visit sites on the opposite extreme. In other words, a regular visitor to a left-wing site such as moveon.org would click over to a right-wing libertarian blog more often than a moderate who just checks CNN for headlines.

But perhaps the most striking finding of the study came in the analysis of real-world communities. Neighborhoods, clubs, friends, work colleagues, family—all these groups proved to be deafening echo chambers compared with all forms of modern media. It turns out that people who spend a lot of time on political sites are as much as three times more likely to encounter diverse perspectives than people who hang out with their friends and colleagues at the bar or by the water cooler. Because we have a natural predisposition to assume that face-to-face social gatherings are intrinsically healthier than mediated ones, the isolation index findings can seem startling. But when you think about it historically, it makes sense. Imagine two stereotypes: an unreconstructed socialist living on the Upper West Side of Manhattan in 1985, and a fervent homophobe living in a small midwestern town around the same time. Both had a vast supply of echo chamber at their disposal back then. Now imagine they time-travel to the present. It is far easier now for each of them to stumble across divergent views— the Manhattan leftie can accidentally follow a link to National Review Online or surf past Sean Hannity on the way to Keith Olbermann, and the midwesterner can come across Ellen DeGeneres talking about her wedding plans, or follow a link to Andrew Sullivan's blog. As David Brooks described it, "This study suggests that Internet users are a bunch of ideological Jack Kerouacs. They're not burrowing down into comforting nests. They're cruising far and wide, looking for adventure, information, combat and arousal."

Like any historic disruption, the transition in journalism—from big media institutions and quasi-monopolies to a more diverse and interconnected peer network—will inevitably be painful to those of us who have, understandably, come to cherish and rely upon the old institutions. The Legrand Stars built by the titans of journalism over the past two centuries established a formidable standard of reporting and analysis that greatly improved on the news systems that had come before them. Given the technological limitations of their time, they were no doubt the most efficient solution to our social need for engaged and accurate news. But we have ample evidence that new models are now possible, thanks both to the technology that makes peer networks possible, and to our growing understanding of how those networks can be harnessed for the greater good. Making the transition to these new models will look like devastation and crisis when viewed from the perspective of the older institutions. But if you believe that in the long run peer networks will outperform top-heavy institutions, the change takes on a different aspect. It starts to look like progress.

What Does
the Internet Want?

A little while ago I published a description of an emerging protest movement that had captured the world's attention with its occupations of urban streets and its vigorous critique of global capitalism:

> It's almost impossible to think of another political movement that generated as much public attention without producing a genuine leader—a Jesse Jackson or César Chávez—if only for the benefit of the television cameras. The images that we associate with the protests are never those of an adoring crowd raising its fists in solidarity with an impassioned speaker on a podium. That is the iconography of an earlier model of protest. What we see again and again with the new wave are images of disparate groups: satirical puppets, black-clad anarchists, sit-ins and performance art—but no leaders. To old-school progressives, the protesters appeared to be headless, out of control, a swarm of small causes with no organizing

principle—and to a certain extent they're right in their assessment. What they fail to recognize is that there can be power and intelligence in a swarm, and if you're trying to do battle against a distributed network like global capitalism, you're better off becoming a distributed network yourself.

No doubt the description is a familiar one, given all the publicity and analysis that the Occupy Wall Street movement generated in late 2011 and early 2012. But here's the catch: I wrote those words in the spring of 2000, more than a decade before the Occupy protesters started camping out in the streets of lower Manhattan. The protests in question were the anti-WTO rallies of Seattle in 1999. I was describing them in the very last pages of my book *Emergence*, which celebrated the power of decentralized networks in many different fields, in nature and in culture: in the distributed intelligence of ant colonies, or city neighborhoods, in the neural networks of the human brain—and, increasingly, on the new platforms of the Internet and the World Wide Web. *Emergence* was not explicitly a political book, but I included the Seattle protests on the last pages as a nod toward a future in which social change would increasingly be shaped by these leaderless networks.

I was not alone in sensing a meaningful connection between the Seattle protesters and the decentralized peer networks of the digital age. Writing in *The Nation* at the time, Naomi Klein had observed, "What emerged on the streets of Seattle and Washington was an activist model that mirrors the organic, interlinked pathways of the Internet." It seemed clear to some of us at that early stage that the model of information sharing that the Internet had popularized was too potent and protean not to spawn offline organizational structures that emulated its core qualities. Seattle seemed just a preview of coming attractions; as the Internet grew to become the dominant communications medium of our age, social movements would increasingly look like the Internet, even when they were chanting slogans in the middle of a city park.

On July 4, 2011, the left-wing Canadian magazine *Adbusters* posted a link to an anti-corporate jeremiad on Twitter: "Dear Americans," they wrote, "this July 4th dream of insurrection against corporate rule." After the link, they included a new hashtag that they also ran on the cover of their print publication: #occupywallstreet. Something about the phrase caught people's attention, and in the ensuing days, the hashtag began to spread across Twitter. A week later, *Adbusters* followed up with an #occupywallstreet tweet that called its followers to action: "Sept 17. Wall St. Bring Tent." If there was ever an object lesson in the power of stacked peer networks, surely this was it: someone tweets a catchy phrase that reverberates as it flows through a social network, and within a matter of months, thousands of people are squatting in downtown streets all across the planet. But that's just one layer. Twitter, of course, was designed as a simple tool for sharing status updates with your friends; the idea of using it to organize a global political movement came from the edges of the network, not the Twitter creators themselves. What's more, the premise of organizing tweets using the hashtag convention was originally proposed and popularized by the users of the service, not the creators. Twitter didn't even formally recognize the hashtag convention for more than a year after users first started tagging their tweets with it. Occupy Wall Street as a meme; Twitter as a political platform; the hashtag as a way of organizing information: all three came from the edges of the network, not the center.

The temptation, of course, is to draw a straight line of techno-determinism between the Seattle protests and the global wave of pro-democratic and egalitarian protest that swept across the planet in 2010 and 2011: from the Arab Spring to the Spanish Revolution to the Occupy movement. The prediction back in 2000 would have gone something like this: because the Internet abhorred hierarchies and top-down command structures, hierarchies and command structures would come under increasing attack, by organizations and movements that looked like Baran Webs and not Legrand Stars. The Internet helped us grasp the real-world potential of peer networks; the next logical step was to take that transformative insight to the streets. This kind of narrative would be both techno-determinist, in that the Internet itself was shaping and cultivating a certain kind of movement, and cyber-utopian, in that the movements that had been created empowered the people against the auto-crats, the 99 percent against the 1 percent.

As tempting as that narrative is, I think it should be re-sisted, or at least complicated. My own experience with *Emergence* should make it clear why. My optimistic predic-tion about the future of decentralized social movements may look prescient now, but that was not always the case. *Emergence*, you see, went on sale the first week of September 2001. I had penned an upbeat account of how decentralized networks could be harnessed to make the world a better place, and the week the book arrived on the shelves, the very city I was living in was attacked by a decentralized net-work of Islamic terrorists who had used the "interlinked pathways of the Internet" to plot and carry out the most deadly attack on U.S. soil in history.

Al Qaeda had its leaders, of course, but as an organization it was always much closer to a decentralized network than a

The peer progressive looks on the War on Terror with a complicated mix of reactions, particularly those revolving around the conundrum of Al Qaeda. The peer progressive believes that the social architecture of the distributed network is fundamentally a force for good in the world, on the order of other related institutions, such as democracies or marketplaces. And the peer progressive believes that the Internet has been the dominant role model and breeding ground for peer networks over the past decade or two. Yet Al Qaeda is self-evidently abhorrent to everything else the peer progressive believes in. It's easy to get misty-eyed hearing stories of Twitter spawning pro-democratic flashmobs in the streets of Cairo; but when those same social architectures are used by reactionary zealots to topple skyscrapers down the street from us, the story gets more complicated.

These complications are closely related to the critiques of cyber-utopianism leveled by writers such as Evgeny Morozov and Malcolm Gladwell. Around the time of the Iranian and Egyptian protests of 2009 and after, Gladwell wrote two much-discussed pieces dismissing the techno-determinist position that the digital peer networks had been instrumental in these popular uprisings, drawing on some of Morozov's argument in his book *The Net Delusion*. "Surely the least interesting fact about [the protests]," he wrote, "is that some of the protesters may (or may not) have at one point or another employed some of the tools of the new media to communicate with one another. Please. People protested and brought down governments before Facebook was invented. They did it before the Internet came along." For Gladwell, getting excited about the fact that the protesters used Twitter to circulate their messages is like getting excited about the fact that they used a lighter, not matches, to set their bonfires alight: what matters is the flame, not the tool you happened to use to start it.

Morozov's work goes even further in its critique of naive cyber-utopianism, outlining the many ways in which the Internet and social networks can be harnessed by authoritarian regimes. Orchestrating your protest movement through the public channels of Facebook may reduce organizational costs, as Clay Shirky would argue, but it also reduces the surveillance costs for the state you're trying to overthrow. The fact that the Internet was built as a peer network, Morozov argues, doesn't preclude autocrats from exploiting its powers for their own gain:

> Perhaps it was a mistake to treat the Internet as a deterministic one-directional force for either global liberation or oppression, for cosmopolitanism or xenophobia. The reality is that the Internet will enable all of these forces—as well as many others—simultaneously. But as far as laws of the Internet go, this is all we know. Which of the numerous forces unleashed by the Web will prevail in a particular social and political context is impossible to tell without first getting a thorough theoretical understanding of that context.

When I think back on the passage I wrote in the months before 9/11, Morozov's critique here echoes in my mind: the "particular social and political context" of that period included the possibility that a deeply reactionary and violent organization would use these decentralized tools to murder almost three thousand people a few blocks from where I lived.

I wrote at the beginning of this book that peer progressives do not consider the Internet a panacea. Does that mean they view it as a purely neutral technology that can be just as easily used for good as for evil, both for building civic participation and propping up dictators? That would arguably be the simplest way to get around the Al Qaeda conundrum: Declare that the Internet is just a tool, and what matters are the values that shape and inform the way you use the tool. After all, peer progressives have a very clear set of values that draw upon the older tradition of progressive politics. They believe in equality, participation, diversity. There's nothing laissez-faire about their agenda for progress. They take the social architecture of the peer network and direct it toward problems that markets have failed to solve.

The problem with the Internet-as-tool argument is that it chips away at that other cornerstone of the peer-progressive worldview: that there are deep-seated forces of progress at work in the world; that the spread of peer networks across the planet is making human society better, on the whole; that history has a direction, just as Hegel and Marx believed, even if the mechanics behind that direction are not exactly what they imagined. Is it possible to believe that the Internet and the Web are pushing us in a positive direction, without becoming naive cyber-utopians? Can the peer progressive believe in the Internet as an engine of progress while still acknowledging the fact of Al Qaeda?

The way you answer that question ultimately comes down to the way you think about the impact of new communications technologies. The basic thrust of all McLuhan-influenced media theory is that each new medium has some deep-seated tendencies, sometimes called "affordances," that shape the messages it conveys in reliable ways. Defined as a tool, television is a communications technology that facilitates the one-way, mass broadcast of sound and moving images. But it has more subtle affordances that appear reliably wherever television is adopted by a wide audience. On the simplest level, television prioritizes visual images and the spoken word over written text, as Neil Postman argued in the influential book *Amusing Ourselves to Death*, written during the heyday of the TV era. You do not need a "thorough theoretical understanding of the context" to assume that the introduction of television will diminish the role of the written word in a given society. And then there are more subtle affordances. For instance, television tends to push political decision making toward the realm of personality and physical appearance. The question of who had the better makeup artist was not a deciding factor in the Lincoln–Douglas debates, nor was it considered relevant whether one of the two candidates was too "stiff" in his presentation, for the simple reason that most voters evaluated the candidates' performances by reading transcripts of their debate. But those are the types of questions that somehow become urgent when political debate unfolds on a television screen.

But something funny happens when you try to detect comparable affordances on the Web. Even seemingly fundamental tendencies are almost impossible to pin down. When the Web first appeared in the early 1990s, it was easy to assume that the medium would counteract the visual culture of television, since the platform relied almost exclusively on pages of hypertext prose. Today there are 4 billion videos streamed every day on YouTube alone. Or consider the echo-chamber argument for micro-customization, all those personalized ads or political messages tailored precisely to your tastes and needs. True enough—just look at the runaway success of Google's AdWords system, which allowed advertising to adapt on the fly to your search queries. But at the same time, the vast majority of information flowing over the Internet's pipes is the least customized form of data imaginable: spam. For every website that knows exactly what you want to read right now, there are probably ten penis-enlargement ads sitting in your in-box.

Is the Internet just more schizophrenic than earlier

forms of media? In a way, yes. The Net is, ultimately, software, and software is all about shape-shifting and simulation. Broadcast television cannot mimic having an online conversation, or playing a video game, or browsing through the aisles of a bookstore. But computers can effortlessly shift between all of those activities—and they can convincingly re-create the experience of watching broadcast television as well. Software interfaces are not fixed properties; they are possibility spaces, open to a near-infinite range of experimentation, which means that the defining affordances of the medium are more elastic than those of traditional media. You can build software that micro-targets individual users, or you can blast out the same penis-enlargement spam to millions. Radio or television or the printed book didn't have that same flexibility. They could tell different kinds of stories or bring to life fictional worlds, but the underlying rules of the medium itself were harder to tweak.

But this capacity for reinvention does not mean the Internet and its descendants are without affordances altogether. In fact, one of the Net's affordances flows directly out of its shape-shifting powers. Because the software networks are more malleable than earlier forms of media, they tend to engage more people in the process of deciding how they should work. In the days of analog telephony or radio, the number of people actively involved in the conversation about how these technologies should work was vanishingly small. If we have too much of anything on the Internet, it's engagement: too many minds pushing the platform in new directions, too many voices arguing about the social and economic consequences of those changes. A medium that displays a capacity for reinvention tends, in the long run at least, to build up a much larger community of people who want to help reinvent it.

That capacity for shape-shifting leads to another affordance: digital networks like the Web can simulate and experiment with different social architectures more easily than other forms of media (and, it would seem, other forms of direct human interaction). Wikipedia turned out to be a wildly successful structure for allowing thousands of contributors to write collaboratively, but there were dozens of other collaborative writing interfaces circulating through the digital world before Wikipedia came along. There are probably more than a hundred start-ups tinkering with the best incentive systems for Kickstarter-style crowdfunding. Some online communities function as virtual anarchies; others build elaborate tools for building reputation and leadership positions. Forty years ago, at the dawn of the Internet era, you could build an alternative social order by retreating to a commune in Mendocino or through the grinding work of changing mass society through protest or public service. But most of the time you had to play by the existing rules. On the Internet, the rules are up for grabs. You can experiment with new forms of governance or collaboration, and when the experiment doesn't generate a useful result, you can move on to the next experiment. That tolerance for risk and failure has two effects: First, the overall system finds its way to useful ideas faster, because the rate of experimentation itself is faster. Second, those ideas can then be ported back into the real world, on the basis of their digital success. The Seattle protesters and OWS modeled themselves on the "organic, interlinked pathways of the Internet," but the movements themselves were populated by people sitting around and talking to one another in a public square.

The capacity for reinvention, for shape-shifting, should not be underestimated. But as an affordance, it is still a bit more meta than we need to be. At the most elemental level, the Internet and its descendants possess this defining property: they make it easier and cheaper to share information. Organizations and individuals and social movements that benefit from the cheap circulation of information—think Wikipedia or open-source software or the spammers—will prosper in that context, all other things being equal. Entities that benefit from scarcity—think local newspapers or state media monopolies—will have a harder time.

Making information cheap and portable sounds like a laudable enough accomplishment, but it can produce surprising outcomes. Like many of my fellow travelers during the early days of the Web, I could sense in the late 1990s that distributed networks were going to empower social movements like the Seattle protests. If someone had been able to show me footage of the Occupy Wall Street protests back when I was writing those lines in *Emergence*, I would have nodded knowingly: *Yep, that's our future.* But those same affordances for cheap and fast information triggered a vast array of outcomes that I failed to anticipate. And not just the Al Qaeda attacks. The Internet may well have made it easier for Occupy Wall Street to form, but it had an even more decisive role in the creation of high-frequency algorithmic trading, which has spawned both immense fortunes and immense instability on Wall Street. Global movements comparable to Occupy Wall Street formed many times before the age of networked computing, as Gladwell observed; they might have had a harder time reaching critical mass without the speed and efficiency of the Net, but they were at least within the realm of possibility. But high-frequency trades are literally impossible to execute in a world without networked computers. If the Internet has a bias toward certain kinds of outcomes, you could make a plausible case that it is more biased toward derivatives traders than it is toward the general assemblies of the Occupy movement.

So what does the Internet want? It wants to lower the costs for creating and sharing information. The notion sounds unimpeachable when you phrase it like that, until you realize all the strange places that kind of affordance ultimately leads to. The Internet wants to breed algorithms that can execute thousands of financial transactions per minute, and it wants to disseminate the #occupywallstreet meme across the planet. The Internet "wants" both the Wall Street tycoons and the popular insurrection at its feet.

Can that strange, contradictory cocktail drive progress on its own? Perhaps—for the simple reason that it democratizes the control of information. When information is expensive and scarce, powerful or wealthy individuals or groups have a disproportionate impact on how that information circulates. But as it gets cheaper and more abundant, the barriers to entry are lowered. This is hardly a new observation, but everything that has happened over the last twenty years has confirmed the basic insight. That democratization has not always led to positive outcomes—think of those spam artists—but there is no contesting the tremendous, orders-of-magnitude increase in the number of people creating and sharing, thanks to the mass adoption of the Internet.

The peer progressive's faith in the positive effects of the Internet rests on this democratic principle: When you give people more control over the flow of information and decision making in their communities, their social health improves—incrementally, in fits and starts, but also inexorably. Yes, when you push the intelligence out to the edges of the network, sometimes individuals or groups abuse those newfound privileges; a world without gatekeepers or planners is noisier and more chaotic. But the same is true of other institutions that have stood the test of time. Democracies on occasion elect charlatans or bigots or imbeciles; markets on occasion erupt in catastrophic bubbles, or choose to direct resources to trivial problems while ignoring the more pressing ones. We accept these imperfections because the alternatives are so much worse. The same is true of the Internet and the peer networks it has inspired. They are not perfect, far from it. But over the long haul, they produce better results than the Legrand Stars that came before them. They're not utopias. They're just leaning that way.

We Have a Winner!

The shrub *Rubia tinctorum*, commonly known as madder, grows delicate yellow flowers in the summer, before small dark red berries begin to line its stems in the early autumn. Belowground, though, lies the plant's true claim to fame: the meter-long roots that contain a molecular compound called alizarin, which absorbs the blue and green wavelengths of light. From at least the time of the Egyptian pharaohs, the roots of the madder plant have been used to produce striking red dyes. Over the centuries, Indian and Turkish dyers refined an elaborate technique that combined madder roots with calf blood and sheep dung, among many other substances, to produce a brilliant hue that came to be known as "Turkey red." By the 1700s, the method had been emulated by European artisans, and madder plants were cultivated extensively in the Low Countries.

Among the well-to-do British population of the eighteenth century, fabric dyed with Turkey red was in high demand. Yet the British Isles lacked a meaningful supply of

Rubia tinctorum in its fields. Although the Industrial Revolution had triggered a historically unprecedented explosion in the textile industry, a key impediment remained for any cloth that needed to be dyed Turkey red: it had to be shipped across the channel to Flanders. With their new steam engines and emerging industrial labor system, the British had revolutionized the production of cloth in a matter of decades. But they still couldn't make that cloth a nice shade of red.

On March 22, 1754, a small group of men gathered at Rawthmell's Coffee-house in Covent Garden and decided to do something about the problem of Turkey red. At this initial meeting, there were eleven of them, a loose network of colleagues and acquaintances, with a strikingly diverse array of professions: naturalists and merchants, watchmakers and surgeons. A drawing teacher and amateur inventor named William Shipley had assembled the group, with the hope of forming a society that would encourage innovation in the arts and in manufacturing. Shipley's idea was that the group would offer awards—called, in the parlance of the day, "premiums"—for solutions to urgent problems that the group itself would identify. Members would "subscribe" to a fund that would allow the society to announce and promote the premiums, and dole out cash awards when the challenges were successfully met. They called themselves the Society for the Encouragement of Arts, Manufacturers and Commerce in Great Britain, though in the early days they were also known as the "Premium Society," given the visibility of the awards they offered. Today the institution, one of the most revered in the United Kingdom, is known as the Royal Society of Arts.

At that initial meeting at Rawthmell's Coffee-house, the eleven men agreed to create two premiums as their inaugural act. Each involved a crucial material for dyeing cloth. One premium offered a reward for the discovery of cobalt ore within the kingdom, thereby facilitating the production of cobalt-blue dyes, and smalt for the creation of blue ceramics. The second premium offered a reward for the cultivation of madder plants on British soil. The premiums were backed by thirty-eight guineas from the initial subscribers, including the Earl of Shaftesbury, and within a few weeks of the initial meeting, an advertisement ran, announcing the awards to the general public. All those who successfully demonstrated that they had produced madder in their fields would be paid five pounds per acre grown.

The Royal Society of Arts would go on to create premiums—and deliver awards—for thousands of innovations: spinning wheels, mechanical telegraphs, naval construction, brocade weaving. But their first act as a group had an almost elemental simplicity to it. They were funding the creation of red and blue.

The madder premium ultimately proved to be an unqualified success. For two decades, the RSA dutifully paid out awards wherever they found that madder plants had taken root. In the end, the RSA disbursed £1,516 in prizes, the equivalent, in today's currency, of more than half a million dollars. By 1775, the society had concluded that the premium had achieved its objectives; thanks to the encouragement of the RSA, the British textile manufacturers no longer needed to export their fabrics to Flanders when they needed them dyed Turkey red. The timing was impeccable. That year, when the British military set sail to subdue the insurrection in the American colonies, their legendary red coats were dyed with alizarin grown on English soil.

The great flowering of mechanical and commercial innovation that took place in England during the eighteenth and nineteenth centuries is conventionally attributed to the entrepreneurial zeal of the inventors and early industrialists who created entirely new markets and new sources of energy because vast economic rewards awaited them if they were successful. But that history cannot be told in its full complexity without reference to organizations such as the RSA, and the tradition of what we now call prize-backed challenges. Countless stories of innovation from the period involve a prize-backed challenge. During the century that followed that first coffeehouse meeting, the RSA alone disbursed what would amount to tens of millions of dollars in today's currency. John Harrison's invention of the chronometer, which revolutionized the commercial and military fleets of the day, was sparked by the celebrated Longitude Prize, offered by the Board of Longitude, a small government body that had formed with the express intent of solving the urgent problem of enabling ships to establish their longitudinal coordinates while at sea. Not all premiums were admirable in their goals: William Bligh's ill-fated expedition to Tahiti aboard the *Bounty* began with an RSA premium awarded to anyone who could successfully transport the versatile breadfruit plant to the West Indies to feed the growing slave populations there. Bligh failed in his first attempt when his crew famously mutinied, but after a second, triumphant voyage, he was awarded the breadfruit prize in 1793. In an ironic twist, the plant readily adapted to the Caribbean climate, but the slaves refused to integrate it into their diets.

In their eighteenth-century form, prize-backed challenges had a nuanced relationship to the free market. On the one hand, they were deliberately conceived as a kind of correction to market failures. They targeted problems that urgently required solutions, but that for whatever reason had failed to elicit a meaningful price from the market in its current state. While the solutions often proved to be a boon for commercial interests once they got into wider circulation, in the short term the market failed to value them properly, so RSA premiums filled that empty space with an "artificial" reward. Without a nation of dyers trained in concocting the intricate formula for Turkey red, growing madder plants on a farm didn't make economic sense. But without a local supply of madder plants, it was hard to build up a labor force that knew how to turn the roots into Turkey red. A premium that made it profitable to plant acres of *Rubia tinctorum* offered a way out of that catch-22.

Like Kickstarter, prize-backed challenges are often responses not so much to outright market failures as they are to market blind spots, or market shortsightedness. Jacob Krupnick may opt to continue funding his video career via the micro-patrons of Kickstarter, but there's little doubt that the success of his seventy-one-minute music video will enable him to follow a more traditional career path—selling his services directly to big media companies—if he chooses. The subscribers funding the RSA premiums were simply perceiving potential value before the private sector managed to. After 1775, the madder plant didn't need the RSA's help; the textile market was more than able to keep the plant in circulation. In 1754, it just needed a kick start.

One other wrinkle complicated the RSA's relationship to

the market. From the very beginning, the society possessed an explicit aversion to patents. The *Rules and Orders* published by the Society in 1765 spelled it out in no uncertain terms: "No person will be admitted a candidate for any premium offered by the Society who has obtained a patent for the exclusive right of making or performing anything for which such premium is offered." Prohibiting patents meant that solutions could circulate more quickly through society, and could be easily improved upon by subsequent inventors or scientists. The RSA wanted you to profit from your idea. They just didn't want you to own it.

The RSA's position on intellectual property was part of a larger intellectual ethos that characterized most Enlightenment science: Ideas improved the more they were allowed to flow through the community; building proprietary walls around those ideas would only retard the march of progress. But those Enlightenment values developed alongside an opposing reaction of sorts in the birth of industrial capitalism, which placed an increasingly heavy emphasis on patents and intellectual property. Over time, the domain of commercial innovation came to be dominated by patent law and the assumption of proprietary ownership, while scientific innovation remained an information commons. By the middle of the nineteenth century, the patent system had become so deeply entwined with the practice of commercial innovation that the RSA repealed their injunction against patents.

Patents—and intellectual property law in general—turn out to be one of the places where the distinction between libertarian and peer-progressive values is most conspicuous. Both peer progressives and libertarians like to see good ideas rewarded with financial gain. Incentives are a core ingredient of a peer network, after all. For the peer progressive, those incentives don't have to be monetary, of course, but there's no doubt the promise of more cash in the bank concentrates the mind wonderfully. If we put a price on ingenious solutions to important problems, we'll have more brains working on those problems, and ultimately more solutions. That much is clear. Matters get more complicated when ownership of those solutions is claimed. If you can prevent anyone else from applying your solution, or improving it, then you as an individual may benefit, but society may suffer because the solution, in its isolated state, stagnates or remains too expensive for most people to enjoy its benefits. Because most libertarians emphasize the fundamental value of private property, the notion of ideas circulating without ownership seems heretical within that framework. As Ayn Rand wrote: "Patents and copyrights are the legal implementation of the base of all property rights: a man's right to the product of his mind."

There is some not insignificant irony here, because libertarians profess to cherish the values of freedom and autonomy more than anything else; capitalism is ultimately superior to socialism because it relies on voluntary, uncoerced relationships between employer and employee, buyer and seller. But all of that emphasis on freedom vanishes when intellectual property is on the table. Good ideas that would naturally flow from mind to mind, and from corporation to corporation, suddenly need to be restrained by the state, unnaturally removed from circulation. (For this reason, a more radical subset of libertarian philosophy actually opposes intellectual property law.) For the peer progressive, however, the conflict does not exist, because the open exchange of ideas is a core attribute of all peer networks. What the peer progressive wants to do is reward people for coming up with good ideas—*and* reward them for sharing those ideas.

Patents conflict with libertarian values on another front as well, because they create Hayekian bottlenecks in the patent review and approval process. Patents have to be granted by the state for them to have any effect, and that involves someone inside the government reviewing the patent application and making an informed judgment. But the United States Patent Office receives roughly half a million patent applications per year, many of them involving arcane pharmaceutical or technological advances that are utterly baffling to a non-specialist. Because the search for "prior art"—earlier instances of the invention that would negate the patent claim—is complex and sometimes tedious, the backlog of patent applications in the U.S. currently stands at close to one million. In Hayek's terms, the knowledge required to evaluate all those proposed patents does not exist in any centralized form but rather in dispersed bits of information scattered throughout the society.

For the peer progressive, one way around that bottleneck is to create reward systems that make patents unnecessary, as in the premiums of the RSA. But where patents turn out to be indispensable, it is also possible to use peer networks to extend and diversify the minds working on the patent review problem. Several years ago, a visionary NYU law professor named Beth Noveck began developing a program that she called peer to patent, which was, in effect, a software platform that allowed outside experts and informed amateurs to contribute to the prior art discovery phase, both through tracking down earlier inventions that might be relevant and explaining those inventions to the overwhelmed examiner in the patent office. Just as Kickstarter widens the network of potential funders for creative work, peer to patent widens the network of discovery and interpretation, bringing in people who do not necessarily have the time or the talent to become full-time examiners but who have a specific form of expertise that makes them helpful to some patent cases. Pilots of peer to patent have now been launched in the U.S., the UK, Japan, Korea, and Australia. Noveck herself went on to oversee the Open Government Initiative in the first years of the Obama administration.

The premiums of the RSA offer another example of how peer networks can productively build on one another via stacks or layers. The overall objectives were discussed and established by a casual network of diverse interests and professions: astronomers, artists, inventors, government officials, chemists, and much more. That diversity meant that the society was more likely to detect important problems that a more unified or specialized group might have missed. Once a premium had been established—and the reward publicly announced—the prize money created a much larger pool of minds working on the problem. John Harrison's story, powerfully recounted in Dava Sobel's bestselling *Longitude*, demonstrates how a prize-backed challenge extends and diversifies the network of potential solutions. Born in West Yorkshire, Harrison was the son of a carpenter, with almost no formal education, and notoriously poor writing skills. When he began working on his first iteration of the chronometer, his social connections to the elites of London were nonexistent. But the £20,000 offered by the Board of Longitude caught the attention of his brilliant mechanical mind. Because the board ultimately doled out small increments of funding with each new draft of the chronometer, Harrison was able to continue tinkering with the problem of longitude for almost half a century. In the language of network theory, the Board of Longitude and the RSA offered prizes in order to exploit the intelligence at the edges of the national network. The eleven men who gathered at Rawthmell's Coffee-house didn't need help getting their innovations into circulation; they were all wealthy or connected enough to ensure that their ideas would get a proper hearing. What they wanted to do was establish a platform that would enable that brilliant but isolated West Yorkshire carpenter to have his say as well.

The rule against patents guaranteed that a third network would open up once a successful idea emerged—because the solution was, by definition, free to circulate through the minds of other citizens who could add incremental tweaks or discover new applications. One network identifies the problems and funds the rewards; another network proposes solutions; and a final network improves on those solutions and puts them to use. Governments and private corporations are not entirely absent from these systems: the Longitude Prize, for instance, was established by government decree, and the madder plants ultimately helped textile manufacturers improve their bottom line. But the vast majority of the important decisions and incentives were not dependent on hierarchical state control or market forces. The prize-backed challenges of the eighteenth century were tremendous engines of progress, but those engines were not powered by kings or captains of industry. They were fueled, instead, by peer networks.

The premiums of the RSA are experiencing a remarkable revival in the digital age. In May 2011, Senator Bernie Sanders of Vermont introduced two bills in the Senate: for the Medical Innovation Prize Fund Act and for the Prize Fund for HIV/AIDS Act. Like the Longitude Prize, the bills proposed by Sanders target a specific problem with vast economic and personal implications, namely, the cost of creating breakthrough pharmaceutical drugs. The problem is an old and vexing one: because new drugs are staggeringly expensive to develop, we have decided as a society to grant Big Pharma patents on those drugs that allow them to sell their innovations without "generic" competition for a period of roughly ten years. Granting those patents may be

good for Pfizer or Merck, but they come with extensive social costs: these potentially lifesaving drugs are far more
expensive during the life of the patent than they would otherwise be, and it's far more difficult for other researchers to
tinker with and improve upon the original innovations.
The result is big profits for Big Pharma, and higher health
insurance premiums for the rest of us. In the most extreme
cases, the patents ensure that patients who need these
drugs simply cannot afford them.

We make these sacrifices ultimately for one reason: Without the incentive of a massive payout from a breakthrough
drug, Big Pharma can't justify the billions of dollars that
can go into the creation of a new product. Pharmaceutical
research and development in this era involve a staggering
amount of risk for a private corporation: you can chase one
promising compound for a decade before finally realizing
that the side effects cannot be overcome, and even when
you do have a product that seems to work, years may pass
before the FDA approves it. That much risk requires a suf-
ficient reward for the R&D to continue. And so as a society
we have decided that it's worth it to grant the artificial monopoly of a patent to ensure that the pipeline of medical
innovation continues.

Patents are so tightly bound up with modern capitalism that you have to take a step back to realize that they

involve the suspension of normal market forces. Markets, after all, rely on competitive pressures between different firms that either bring prices down or encourage the development of new products. But when the state gives firms a patent, it effectively grants them the right to opt out of that competition. If you think capitalism is purely about motivating people with financial rewards, then patents make sense. But if you think capitalism is about the creative power of competition, then patents are a betrayal of that system.

For a peer progressive, the existing compromise around pharmaceutical drugs is objectionable on multiple levels. In a strange sense, it is both too dependent on the marketplace and not dependent enough. It assumes that the rewards that encourage innovation should come exclusively from consumers buying the ultimate product, and at the same time, it suspends the laws of competition to ensure that those rewards are as large as possible. By suspending those laws and creating artificial monopolies, pharmaceutical patents eliminate network density and diversity. New ideas are locked down inside a single organization and cannot circulate through a wider network—at least until the patent expires. And the dependence on consumers' paying for the drugs means that Big Pharma steers toward drugs that are of interest to wealthier patients, and away from drugs that might solve life-or-death crises for poorer segments of the population. In other words, R&D money chases the next Viagra and not a malaria vaccine.

The Sanders pharmaceutical prizes can be seen as a peer-progressive alternative to the existing patent-based solution to medical innovation. The proposed bills are complicated, but at the most elemental level they do two things: They offer billions of dollars of prize money for new pharmaceutical innovations. And they mandate that the prizewinners share their innovations open-source style, and forgo any attempt to patent their discoveries. The approach closely resembles the original premiums of the Royal Society of Arts. By creating an outlandish award for a successful product—a successful treatment for HIV/AIDS could garner billions of dollars in award money—Sanders seeks to increase the network of organizations attacking the problem. And by mandating that the innovations not be shackled to the artificial monopolies of patents, these bills increase the network of people who can enhance and refine those innovations. And of course, the bills ensure that breakthrough drugs arrive on the market at generic prices, which benefits both consumers and the health insurance industry.

Here again, the peer-progressive approach breaks from the mirrored alternatives of Big Capitalism and Big Government. The Sanders bills are not examples of the state itself trying to come up with breakthrough drugs, or even making direct decisions on which firms to back in pursuing those goals. And it's certainly not the same old solution of letting the market solve the problem on its own. Instead, what the bills try to do is use government dollars—and publicity—to widen the network involved in solving these crucial problems, and to make it easier to share and stack the solutions that emerge. Part of the beauty of the approach, in fact, is that it doesn't rely exclusively on taxpayer money: in the Sanders plan, the health insurance industry contributes a significant portion of the prize money, a pittance compared with the vast fortune the industry would save in a world where new drugs appear on the market at generic prices.

The proposed legislation describes its approach as "delinking research and development incentives from product prices." In other words, you don't have to charge a king's ransom for your drug if there's an actual king's ransom, in the form of a state-funded prize, waiting for you once you bring the drug to market. The bills also carve out additional prize money for intermediary tools and practices that widen the diversity and density of the research network. Each year, billions of dollars would be available to institutions that release their findings into the public domain, or at least grant royalty-free open access to their patented material. The Sanders bills set incentives that reward not just finished products but also the processes that lead to breakthrough ideas. They make open collaboration pay.

Bernie Sanders's colleagues in the Senate may not be ready to grasp the creative value of prize-backed challenges, but RSA-style premiums are proliferating throughout the government, on both the federal and the local level. Software-based competitions—such as Apps for America or New York City's BigApps competition—reward programmers and information architects who create useful applications that share or explain the vast trove of government data. The Obama administration's Open Government Initiative specifically directed government agencies to create prize-backed challenges to encourage innovation in all sectors. In concert with a for-profit start-up named ChallengePost, the U.S. General Services Administration launched a website in 2010 called Challenge.gov, which weaves together all the prize-backed challenges currently available throughout the government. The site currently lists hundreds of active challenges, everything from developing new fuel scrubbers for the Air Force to a Healthy Apps contest sponsored by the Surgeon General's Office.

For years, military security forces and civilian police have wrestled with the challenge of stopping fleeing vehicles without harming either the occupants of the vehicle or the security forces themselves. High-speed chases were notoriously ineffective and often resulted in collateral civilian injuries; tire shredding strips will stop a runaway vehicle, but they need to be installed beforehand and thus are useless in most cases. Hoping to find a solution to this vexing problem, the Air Force Research Laboratory launched a prize-backed challenge, inviting minds from around the world to propose innovative solutions. Within sixty days they had a winner, submitted by a sixty-six-year-old mechanical engineer who lived in Lima, Peru. The engineer had designed a remote-controlled vehicle that accelerated under the fleeing car and then released a massive air bag that lifted the car's tires off the road, causing it to slide safely to a stop.

Because they are targeted explicitly at individuals or groups who are not on the government's payroll, state-funded challenges offer a way to route around Hayek's bottleneck. The problems may be defined by government insiders or bureaucrats, but the solutions arise from the edges of the network, not the center. They widen and diversify the web of collaboration, encouraging the John Harrisons of the world to make a contribution, despite the fact that they have no direct connection to the authorities in Washington.

The administration has taken to quoting Sun Microsystems cofounder Bill Joy: "No matter who you are, most of the smartest people work for someone else." Prize-backed challenges are an acknowledgment that governments work better when they tap the intelligence of the wider population. The power of these challenges does not merely arise from the sheer number of people involved but also from the intellectual diversity of the population. NASA cosponsored a challenge to elicit algorithms that would help computers map the dark matter in the universe. They created a "live leaderboard" of entries, so that participants could learn from their competitors and refine their submissions over time. Leading algorithms were submitted by a Ph.D. student researching satellite photos of glaciers, a neuroscientist at Harvard Medical School, and a signature verification expert from Qatar University.

The most celebrated prize-backed challenge of the Obama administration, however, predated the Challenge .gov site: the Race to the Top competition, sponsored by the Department of Education as part of the 2009 stimulus package. The idea for Race to the Top had begun almost a decade before, when the Clinton education adviser Jon Schnur began systematically tracking the strategies and best practices used by high-performing schools and districts that served low-income neighborhoods around the country. Schnur's organization—New Leaders for New Schools— had recognized that the secret to education success did not lie in just a single charismatic principal or superintendent but rather in a remarkably consistent series of reforms, including consistent teacher assessments and training programs; clear pathways for the promotion of aspiring teachers and aspiring principals; or the district-wide introduction of more experimental charter schools. Schnur was looking, in effect, for positive deviance in schools in low-income areas.

Schnur's findings had influenced the education policies of Obama as a candidate, so when Arne Duncan, Obama's education secretary, started thinking about the education component of the stimulus act in the weeks after the inauguration, Schnur and Duncan began brainstorming for ways to encourage more schools to adopt the reforms that Schnur's team had found to be the most effective. They wanted an approach that would steer the states toward a series of broad goals, without the federal government's having to micromanage the individual steps taken. They also wanted to avoid the appearance of Washington's issuing a top-down mandate that deprived local school systems of the ability to make their own decisions. So they hit upon the idea of a competition among the states—a competition with $5 billion of prize money. They took the examples of positive deviance from Schnur's older study and created a list of statewide goals with a specific number of points associated with each goal. "Ensuring successful conditions for high-performing charters" was worth 40 points; "improving the effectiveness of teacher and principal preparation programs" was worth 14. States were encouraged to submit proposals that demonstrated their plans for achieving those goals, with the promise that the $5 billion would be doled out among the states with the highest-scoring plans.

The beauty of a prize-backed challenge like Race to the Top, or the premiums of the RSA, lies not just in the incentive it creates for the ultimate winners but also in the way it spurs activity among individuals or groups who end up not winning the competition. In the end, forty-six states submitted Race to the Top proposals, and while twelve of them (led by Florida and New York) ultimately took home prize money, many of the reforms proposed by the "losers" were ultimately enacted. The allure of the competition itself generated tremendous publicity, on both national and local levels, which also helped promote the reforms Schnur and Duncan were championing. In effect, the prize-backed challenge approach greatly increased the productivity of the taxpayer dollars spent: by promoting change in school districts that ultimately didn't receive a dime of new funding, and through the free publicity generated by the competition itself. The $5 billion was slightly more than 1 percent of the overall education budget, yet Race to the Top has generated far more attention than any other Obama education initiative to date.

Like Bernie Sanders's medical-innovation bills, Race to the Top was an attempt to create market-style rewards and competition in an environment that was far removed from the traditional selection pressures of capitalism. While the federal government served as the ultimate judge of the competition (effectively playing the role of the consumer in a traditional marketplace), both the problems and the proposed solutions emerged from the wider network of state and local school systems. Yes, Schnur and Duncan were using the power of the federal government to set goals, but the goals themselves had been established by studying the local strategies employed by principals and school superintendents all around the country. And the decisions about how to best achieve those goals also resided at the edges of the network, not in the White House. With Race to the Top, the federal government's role was merely to amplify and promote the innovative strategies that had already emerged on a local level.

A prize-backed challenge is a way of steering a peer network toward a goal, without restricting the route the network chooses to get there. In theory at least, the Board of Longitude didn't care which method ultimately allowed ship captains to determine their longitude at sea. They just wanted the problem solved. (In practice, it should be said, the board ultimately developed a bias toward astronomical solutions—which, many historians think, wrongfully delayed Harrison's award for his clock.) The prizes and premiums give the swarm a target to run toward; if some of them take elliptical loops to get there, so much the better. Some of the best ideas come from unlikely swerves.

The question, of course, is who gets to define the goal. Markets work so well in the long run because both the goals and the routes are defined by peer networks. Customers decide what their priorities are by buying certain things and not others: they start buying smartphones or Turkey-red drapes, and the network of producers starts experimenting with new ways to satisfy those needs. In the artificial market of the prize-backed challenge, the goal has to be defined in some other way. Prize-backed challenges issued by a dictator or a smoke-filled room of experts might well spur innovation at the edges of the network, but the goal itself would emanate from the core of a Legrand Star. That makes those goals vulnerable to the whims of an individual tyrant, or the blurred vision of isolated bureaucrats, or the echo chamber of groupthink.

Those problems disappear, however, if the goals are defined by a peer network as well. That's exactly the approach taken by the most celebrated sponsor of prize-backed challenges in the modern age: the X Prize Foundation. Now the source of dozens of million-dollar prizes in a wide range of fields, the organization took its original inspiration from the Orteig Prize won by Charles Lindbergh in 1927 for the first successful transatlantic flight. The X Prizes began in the mid-1990s, when the aerospace engineer and entrepreneur Peter H. Diamandis announced a competition that would spur innovation in the then nonexistent private-spaceflight industry. Ten million dollars would be awarded to any group that could carry three people beyond the earth's atmosphere, approximately sixty-two miles above the surface of the planet. (To prove that the solution was a durable one, the prizewinner had to complete the mission successfully twice in two weeks.) Eight years after Diamandis announced the competition, the Ansari X Prize was awarded to the creators of SpaceShipOne, led by aerospace legend Burt Rutan and Microsoft cofounder Paul Allen. Not only did SpaceShipOne break the government monopoly on space travel, it has also helped spur almost $2 billion in public and private funding for the civilian spaceflight industry, just as Lindbergh's flight helped usher in the age of commercial air travel.

Today the X Prize Foundation offers more than $100 million in prize-backed challenges that reward innovation in genomics, personal health care technology, automobile energy efficiency, and oil cleanup. Google has cosponsored a $30 million Lunar X Prize for the first group that can successfully land a robot on the moon. The foundation has dozens of prizes in development as well, some of them focused on attention-grabbing feats like space or deep-ocean exploration, but others rewarding breakthroughs in education and the life sciences. The foundation prides itself on the way it develops its new prizes, drawing on a peer network of diverse interests to "ensure the input of a variety of perspectives." The network of advisers that created the Lunar X Prize included Internet entrepreneurs, technology historians, NASA officials, MIT aeronautics professors, and pioneers in the private-spaceflight industry. Each prize emerges out of an intense, layered process: researching the field and defining worthy problems, and then crystallizing the best set of objectives for the prize that will capture the attention of would-be winners as well as the wider public.

The X Prize founders may have been inspired by the Orteig Prize and the *Spirit of St. Louis*, but the peer networks they have formed are truly the descendants of those eleven men in that Covent Garden coffeehouse, trying to bring more madder plants to the British Isles. The problems have changed, but the fundamental approach remains the same: a diverse network working outside the marketplace establishes a worthy goal, and an even more diverse network sets out to find a way to reach it—even if the goal in question turns out to be more than 200,000 miles away on the surface of the moon.

Liquid Democracies

The city of Porto Alegre lies at the edge of an immense freshwater lagoon, near the southern border of Brazil. Overshadowed by its more glamorous and cosmopolitan neighbors to the north, Rio and São Paulo, Porto Alegre has quietly grown into the fourth-largest metropolitan area in Brazil, with more than 4 million residents. Much of that growth came in the second half of the twentieth century: Porto Alegre nearly quadrupled in size between 1950 and 1980. That kind of explosive growth can be devastating to a city; infrastructure can't keep pace with the population, which creates large pockets of underserved communities lacking basic human necessities. By the late 1980s, Porto Alegre's growth had left it with vast favelas (or shanty-towns) where close to a million people lived in improvised shacks, almost all lacking clean drinking water or waste-removal systems. (In 1989, for instance, less than 50 percent of the city's population had sewer access.) Corruption and incompetence in the city government—as well as a lack of representation on behalf of the favelas—meant that the existing resources were rarely applied to the communities that needed them the most.

In the late 1980s, as top-heavy communist governments were imploding around the world, the citizens of Porto Alegre began an experiment in local governance that was in many ways more true to the original ethos of socialism than the autocratic regimes that had come to embody it. The Brazilians called it *orçamento participativo*, or participatory budgeting. Instead of handing off the decision of how to spend taxpayer dollars to the elected officials, the citizens of Porto Alegre decided to participate in that decision-making process directly. It was their money, after all; why shouldn't they have a direct say in where it went?

Participatory budgeting involves multiple steps of civic engagement. In April of each year, the sixteen regions of Porto Alegre conduct general assemblies, at which the results of last year's investments are reviewed and new projects contemplated. A series of smaller neighborhood meetings follows the general assemblies, and here locals can debate the priorities for the new year's budgets. In June, thematic groups are formed to discuss specific areas of concern: sanitation, zoning, health care, and so on. Each region then consolidates the suggestions from the neighborhood groups into a unified ranking of priorities for the budget. A citywide Municipal Budget Council, with elected representatives from each region, then allocates resources based directly on the regional requests. Each region receives funding based on two primary criteria: the overall population, and the existing quality of infrastructure and other civic resources in the community. The poorer the region, the larger its slice of the annual budget. While there are elected officials on the Municipal Budget Council, they are not, technically, deciding where the taxpayer dollars should go. Instead, they are simply routing money to the projects and services that the regions have collectively chosen as priorities. The money comes from the state. But the decisions come from the streets.

The institution of the *orçamento participativo* transformed Porto Alegre almost immediately. Within seven years, almost 95 percent of the city had access to the sewer system. Before 1989, only 2.5 miles of new roads were paved each year; after the new budgeting system was put in place, that number jumped to more than twelve miles per year. The transparency of the process meant that corruption and waste vanished almost overnight. As the Brazilian political scientist Rebecca Abers writes, "It was impossible for money to disappear, for contracts to be overpriced, for promises to be ignored, and for unnecessary investments to be made." Allocating the funds to the neediest communities meant that the favelas saw a tremendous improvement in quality of life. But it also had a more subtle but equally important effect: it demonstrated that civic engagement in those communities could lead to tangible results.

Since Porto Alegre's first experiments with participatory budgeting, the practice has spread to local governments around the world. From 2000 to 2005, the number of participatory budgets in Europe spread from six to fifty-five; roughly 10 percent of municipal budgets in Spain are now allocated on the basis of Porto Alegre–style civic participation. *Orçamento participativo* has even spread to a few cities in the United States, including districts of Chicago and Brooklyn, where discretionary budgets are now divvied up on the basis of citizen feedback. In the United States, the recent talk about reinventing government has focused on the potential breakthroughs that Internet-based engagement can produce. But the history of those Porto Alegre general assemblies suggests that the more radical advance could well come from the simple act of neighbors gathering in a meeting hall or a church or a living room, and drafting up a list of the community's needs.

There are few institutions of modern life that are more universally respected than the institution of representative democracy. You hear a lot of complaints about the people who happen to populate the institution, but the tradition of democracy itself—its social architecture—is generally sacrosanct. Yet by the standards of existing peer networks, representative democracies are more centralized and top-heavy in their day-to-day operation. The leaders themselves are selected from below, but they govern from above. The voters choose the "deciders," to borrow a term from a recent political leader; they don't make policy decisions themselves. The American Founders had endless debates about the right balance between federal and state authority, but they were united in the belief that direct democracy would be a mistake. For the most part, that assumption has remained in place for more than two centuries. The few instances in which direct democracy has erupted—most notoriously in California's proposition system—are generally considered to be disasters.

The hierarchical nature of most governments is not just a theoretical test for peer-progressive values. The fact that these institutions remain Legrand Stars is directly related to one of the most depressing trends over the past decade: the growing contempt for public officials. There may be numerous unsung examples of incremental progress around us—think of the airline safety and crime statistics we began with—but there is no denying that the narrative of government trust is a profoundly bleak one.

Why have those attitudes departed so dramatically from

the steady improvements we have seen in so many other sectors? For a peer progressive, part of the answer involves the consolidation of decision-making power into smaller and less diverse groups.

Begin with the money. The staggering cost of running a political campaign has introduced a crippling distortion into our legislative priorities. The average congressperson now spends more than 30 percent of his or her time fundraising, and naturally it makes sense to focus those efforts on the individuals or organizations that can spend the most money. Attracting that money—and then raising it again during their tenure in office for the next campaign—inevitably draws politicians toward the priorities and values of the contributors. That might not be a bad feedback mechanism with a diverse network of potential funders. (The whole point of a democracy, of course, is that the politicians will be drawn toward the positions of their backers at the ballot box.) But in practice, politicians have to focus on the big donors, and the big donors represent a tiny fraction of the wider electorate. Harvard's Lawrence Lessig has documented in his essential recent work on campaign-finance reform that the part of the population that maxes out its political campaign donation is .05 percent—a group that makes Occupy Wall Street's 1 percent look like the proletariat. The result is that our political leaders spend time on issues that are less vital to the broader electorate, because they are steered toward those issues by the subtle influence of their big funders. And in many cases, they take the wrong side on those issues, or at least ally with the .05 percent over the rest of us. Their primary constituents become funders, not voters.

As Madison wrote in No. 39 of *The Federalist Papers*, "It is ESSENTIAL to such a government that it be derived from the great body of the society, not from an inconsiderable proportion, or a favored class of it; otherwise a handful of tyrannical nobles, exercising their oppressions by a delegation of their powers, might aspire to the rank of republicans, and claim for their government the honorable title of republic." The tyrannical nobles of today are not members of an Old World aristocracy; they are the hedge fund managers and teachers' unions and Big Pharma—the individuals and organizations whose financial power endows them with a staggeringly disproportionate influence over the legislative agenda of our political leaders.

Consider just this one example, among countless potential others. Over the past two decades, as Lessig notes, sixteen different new laws have passed that add additional teeth to copyright restrictions, while not a single one has passed restricting industrial carbon output. There are genuine debates to be had over music piracy and global warming, but surely the vast majority of us would agree that factories pumping carbon into the atmosphere with impunity are a more pressing concern than illegal Kanye West downloads. Yet the congressional record is clear: sixteen to zero. Why? Because content businesses and carbon polluters spent billions of dollars supporting candidates from both parties over that period. So we have a political class that spends its time protecting Mickey Mouse and Lady Gaga instead of the planet.

A representative democracy in which a microscopic portion of the overall population gets to pass the laws of the land, and in which their decisions are heavily weighted by a small coterie of powerful individuals and organizations— this is a system almost fiendishly designed to offend both the Founding Fathers and the Hayek disciples. It's a betrayal of democracy in that the campaign donations matter almost as much as the votes themselves, and it's a betrayal of Hayek's principles in that a small number of elites get to define the problems and solutions for an increasingly complex society. The modern system of campaign funding is an equal-opportunity offender.

The peer-progressive response to this travesty is simple enough in theory. Don't eliminate campaign financing altogether; diversify and decentralize it. A hundred years ago, during the last great crisis of political accountability, the political scientist Robert Brooks observed in his classic study of political corruption: "It is highly improbable that the question of campaign funds would ever have been raised in American politics if party contributions were habitually made by a large number of persons each giving a relatively small amount."

"A large number of persons each giving a relatively small amount." We have seen this formula before, in the micro-patronage of Kickstarter and other crowdfunding services. To a certain extent, the Kickstarter model is already alive and well in the world of campaign finance: 90 percent of the donations to President Obama's 2008 campaign came from individuals giving $200 or less. But as we have seen with the super PACs of the 2012 election campaign, wealthy individuals and organizations continue to have a wildly disproportionate impact on election cycles, and ultimately on the legislation proposed (or ignored) once the candidates make it into office.

Building on the work of Yale's Bruce Ackerman and Ian Ayres, Lessig himself has proposed an ingenious solution to this crisis, which entails creating a Kickstarter-style approach for the public funding of campaigns. Lessig's scheme would create what he calls "democracy vouchers." Taxpayers would be allowed to contribute fifty of their tax dollars to support the campaign expenses of a specific candidate, or a group of candidates. They would also be allowed to supplement those contributions with an additional hundred dollars of funds that would not come from their tax bill. Taxpayers who don't earmark their fifty dollars for a candidate would see their tax dollars go to the political party with which they are registered. The funds from taxpayers who don't specify a candidate and who are not affiliated with a party would support the infrastructure of political elections: voting centers, debates, and so on. The beauty of this approach is that it ensures that taxpayers aren't inadvertently funding politicians who are on the opposite side of the political spectrum. You can back your favorite candidates or your party, but if you don't explicitly choose a side, your fifty dollars supports the institution of democracy itself.

The candidates would have access to this pool of money under one condition: They would have to forswear any other source of funding. The pool would be big enough that it would make the existing system—courting wealthy donors via endless, distracting fund-raising events and outreach—seem like a colossal waste of time. Lessig estimates it would make $6 billion in funds available to candidates and parties during each two-year election cycle, more than the combined total contributed to congressional candidates and the political parties in 2010. "As a candidate," Lessig writes, "you would not have to starve to be good." By making it financially competitive, the "democracy vouchers" approach would not require a constitutional amendment to ensure its success, since the system itself would be voluntary. A candidate would always be free to rely on big donors and special interests, and turn down the voucher funding. But in practice, the $6 billion would be hard to resist.

Democracy vouchers strike me as being a perfect example of peer-progressive values at work. It is not a brute-force hijacking of campaign finance by the state. The decision of where the dollars should flow is not made by the bureaucrats at the center of a Legrand Star, extracting taxes from Tea Party members and routing them to pay for the Chardonnay at Nancy Pelosi's reelection soirees. The $6 billion gets divvied up at the edges of the network, through millions of individual decisions made by the taxpayers themselves, just like the participatory budgets of Porto Alegre. Engaged individuals would still be able to support their candidates financially, albeit with a finite amount of money. Democracy vouchers would decentralize and diversify the sources of campaign funding, creating a true market for campaign funding, not one dominated by a minuscule fraction of the electorate. In this respect, they are true to libertarian principles. Yet democracy vouchers also acknowledge that purely laissez-faire campaign finance leads ultimately to a betrayal of the republic itself. If we simply let the richest individuals and organizations dominate the funding cycle, a different kind of Legrand Star emerges, in which .05 percent of the population controls the country's legislative priorities. If democracy vouchers were implemented, the gravitational pull of the big donors would disappear overnight; legislation would once again reflect the interests and needs of constituents, not funders. Madison's "tyrannical nobles" would lose their influence, and the politicians would be free to spend their hours solving our country's problems, instead of dialing for dollars.

Could peer networks also transform the way government works, and not just the way campaigns are financed? Here we venture into more complicated terrain. It seems reasonable enough to assume that the American Founders would be appalled by the way campaign finance has come to distort and distract today's political leaders. Lessig himself would be the first to admit that his democracy vouchers are only one of many potential approaches that could improve campaign finance by making it more like a peer network. But tinkering with the way laws get proposed and passed raises the stakes significantly. The United States is a republic, not a direct democracy; we elect what we think are the best leaders, and the leaders decide what's best for us. If the Founders agreed on anything, it was that direct democracy (Madison called it "pure" democracy) was almost as bad as its opposites: tyrannies, monarchies, theocracies. As Madison put it in *Federalist* No. 10:

> [In a pure democracy,] a common passion or interest will, in almost every case, be felt by a majority of the whole; a communication and concert result from the form of government itself; and there is nothing to check the inducements to sacrifice the weaker party or an obnoxious individual. Hence it is that such democracies have ever been spectacles of turbulence and contention; have ever been found incompatible with personal security or the rights of property; and have in general been as short in their lives as they have been violent in their deaths. Theoretic politicians, who have patronized this species of government, have erroneously supposed that by reducing mankind to a perfect equality in their political rights, they would, at the same time, be perfectly equalized and assimilated in their possessions, their opinions, and their passions.

The Founders took the threat of tyrannical majorities very seriously. In the system proposed in the U.S. Constitution, the people are sovereign, but the sovereign has to be protected from its own excesses: the herd mentalities and the subtle (or not so subtle) repressions of minority opinions that inevitably arise when the intermediaries are taken out of the mix. So voters don't propose or vote on legislation directly—unless you count the ballot initiatives that have drawn so much criticism over the past decade or two. The voters choose the lawmakers, but the lawmakers make the laws.

But direct democracies have another flaw, one that ultimately involves the same information overload that Hayek detected in central planning: there's just too damn much data out there. We live in enormously complex societies; building public policy that does justice to that complexity requires expertise in economics, law, sociology, and increasingly technology and science. It requires intense study and debate. Hayek's point was that the complexity ultimately made it impossible to funnel all that information into a narrow enough channel to make the signal meaningful—or legible, to use a term from James Scott's *Seeing Like a State*—to the central planners. But everything that Hayek thought was wrong with central planning is even more egregious in direct democracies. Voters don't want to spend their time mastering the subtleties of tort reform or agricultural subsidies. They want to spend their time building companies or playing ukulele or hanging out with their kids. They want the pursuit of happiness. They don't want to feel like they have to become experts in derivatives regulation to be engaged citizens. At least for the central planners, central planning is their day job. The rest of us don't have that luxury.

Yet there is another funnel at work here, and that is the one that compresses the voter's complexity down to a single candidate—or, as it generally works in practice, to a single party. You might like one candidate over her rivals, but you don't necessarily agree with her on all the issues. But you don't get to vote for issues; you vote for candidates. So you shave off the edges of your beliefs and vote for the closest match. This book is itself a kind of extended complaint about how that funnel limits our options. Peer progressives have beliefs that don't fit easily into existing political categories, but when we walk into that voting booth, we're forced to compress those beliefs down into the predictable patterns that the two parties offer, in the United States at least. To give only the most obvious example, peer progressives have a constitutive wariness of big corporations and big unions. What's the political party that matches that sensibility? It's a sign of how restrictive the funnel is that the closest match is probably the Tea Party, which (1) is not even a proper party, and (2) has almost nothing else in common with the peer-progressive agenda.

Direct democracies do away with these constraints. You

don't have to mainstream your opinions at the ballot box. When the legislation comes up that proposes an end to teacher tenure, you can vote for it directly, even though you're also supporting an initiative that caps executive pay at Fortune 500 companies. Every legislative session is a Chinese menu. You make the laws you want, or at least you get to vote for them directly. Your unique political worldview may be expressed through the totality of those gestures, and not through the cardboard figureheads of your "elected representatives."

So this is a tale of two funnels. The world is too complex to compress down into a package that would be understandable to the average voter, which makes direct democracy a pipe dream. But the voters themselves have a complexity to their political values that can't be expressed in individual votes for candidates or parties. But are these two funnels inevitable? Could we imagine another approach that somehow allows the engagement of direct democracy without the information overload?

For the last thirty years of his life, the English mathematician and author Charles Lutwidge Dodgson (more widely known by his pen name, Lewis Carroll) served on the governing board of Christ Church College at Oxford, where he had studied as an undergraduate. The experience immersed Dodgson in the political battles that flourish in any university: student-body elections, debates over the construction of new facilities, the appointment of new professors and fellows. During the 1870s, Dodgson wrote a series of pamphlets and letters discussing his emerging theories about group decision making and voting schemes. His interests ran parallel with national currents: by the 1880s, England had embarked on a long and sometimes tortured process of reforming the parliamentary system, which ultimately resulted in the reform acts of 1884 and 1885, which greatly extended the number of Britons who were allowed to elect members of the House of Commons.

Dodgson's playful but mathematical genius was naturally drawn to these sorts of problems: how to design the "game" of the electoral system to ensure results that best represented the will of the people. In 1884, just as the reform acts were making their way through Parliament, Dodgson published a short pamphlet called "The Principles of Parliamentary Representation." The title suggested a work of political agitation, but the essay itself read like a game-theory treatise, complete with mathematical tables and formulas. Dodgson's contributions to the debate over electoral reform were largely ignored at the time, but the essay contained the seed of a simple idea that began to germinate over the following decades: elections would treat votes as a kind of currency; you could "spend" your vote on one candidate, who would then be free to "spend" the vote he had received from you on another candidate. You could, in a sense, pay your vote forward. Dodgson's system was designed to deal with a problem specific to the British parliamentary system: what to do when the first run of votes didn't elect (or "return," as the British say) enough members to fill the seats for a given district. In that scenario, Dodgson suggested, the candidates could take the votes they had received from the electorate, and "spend" them on themselves or on other candidates to fill the remaining seats. Dodgson proposed that candidates be allowed to "club together" their votes to put one candidate over the threshold, forming alliances whose power would be determined by the original votes submitted by the electors. The candidates would make the decision of who should fill the remaining seats, but their relative power in making that decision would be determined by the overall number of votes they had received from the voters themselves.

More than a century after he published his pamphlet, Dodgson's system has come to play a central role in some of the most interesting and provocative new thinking about alternative models of democracy. The approach is sometimes called "proxy" or "delegate" voting, but the term that has seen the widest adoption is the more evocative one: "liquid democracy." In a liquid democracy, the line between voter and candidate is much more fluid than in traditional democracy, precisely because votes can flow through the broader network of voters. Let's say it's November of a presidential election year, and you have strong feelings about who should be the next president, and almost as strong feelings about who should represent your state in the Senate. But the thirty other names on the ballot—the city council election, the state senate race, the attorney general—all these people are total strangers to you. They're strangers because of the information overload funnel; you don't have the time to follow all these different races. (It was hard enough just to carve out time for the presidential debates.) If you happen to belong to a political party, you can always just vote across the board for that party, but then that's the other funnel at work, compressing your idiosyncratic values down into a legible straight line.

But a liquid democracy gives you a way around those two funnels. Let's say you have a friend who is intensely involved with the public school system, and whose opinions you have always respected. In a liquid democracy, you can transfer your vote to your friend, and authorize her to "spend" it as she sees fit. You retain control of the votes that you care about, or the votes where you feel you have enough expertise to make an informed decision. But for the races that you don't feel as strongly about, you can effectively offload your vote to someone who has more information than you do. Your friend the public school expert can pool together dozens or hundreds—even thousands—of votes if she wants; she can use her expertise to build up more influence over the outcome of the election that she cares the most about. Or she can pass along the bundle of votes she's collected to some higher authority, who then represents the entire ensemble. As the votes circulate through the peer network of voters, expertise or wisdom becomes more heavily weighted in the final election results. By collecting additional proxy votes from her extended network, the public school expert has a louder voice in determining who should be elected superintendent than the twenty-something who couldn't care less about local schools. But that influence was earned, not bought. The proxy voter who has a thousand votes to spend didn't get that disproportionate influence by making extravagant campaign donations, or running super PAC–funded attack ads, or lobbying on K Street. Her influence flows from the respect of her peers.

Those proxy votes could be bought, of course. A phony public school expert could walk through a neighborhood handing out twenty-dollar bills to anyone willing to pledge his school superintendent vote to her. But this is true of any democracy. We have voter fraud laws for a reason. In our current system, the aspiring superintendent could buy support directly from the voters, assuming he didn't get caught doing it. The difference is that in a liquid democracy, the proxy voter doesn't have anywhere near the same incentive to cheat the system. If the superintendent wins the election, he gets to be superintendent, after all. The proxy voter just gets to see her favorite candidate win.

Liquid democracies artfully avoid the twin funnels of direct democracies and republics. By transferring your vote to your more knowledgeable friend, you've weakened the funnel of simplified, party-line voting: your proxy voter might well support candidates from other parties if she thinks they're the most qualified. But at the same time, you've also overcome the information-overload funnel. You don't need to be an expert in everything for your vote to matter. You can pick your targets, and let the people you trust in other fields make the remaining calls.

A democratic system where you don't have to master the full complexity of society to cast an informed vote, a system where you can transfer small bits of your sovereignty to your peers—in these kinds of systems, new forms of direct engagement with governance become possible. Liquid democracies open up the possibility for popular votes on new laws or other state interventions, precisely because they don't demand general expertise from every individual voter.

Imagine a city government that is debating whether to

build a sports stadium at a busy intersection. The project is controversial on a number of fronts: it promises to create construction jobs, and a long-term economic benefit for the city from the crowds that will converge on the stadium on game days. But the costs of traffic congestion and eminent domain evictions are real ones as well.

These kinds of decisions are the very essence of politics, involving competing priorities, values, and incentives—as well as competing predictions about the future. Local residents don't want to see their quiet neighborhood streets overrun with drunken sports fans on game nights. Sports fans want a sleek new stadium. Local unions want the construction jobs. And each constituency has a different prediction about how the stadium will turn out if it is in fact built.

We created republics to adjudicate these conflicts. But perhaps peer networks could do a better job of resolving them.

In a traditional democracy, the stadium decision would play out along these lines: First, the prospect of the stadium might influence the outcome of a local election, if it became a big enough issue. A mayor might take office because of his support (or lack thereof) for the stadium project. After that vote, though, the decision would ultimately come down to the representatives. Hearings would be held to allow the various groups to express their opinions; the voice of the people would be heard (distantly); and then the politicians would decide. If they made a terrible decision, one totally out of sync with the popular will, there would be the ultimate feedback mechanism of getting blasted out of office a year or two later.

But imagine this decision is made instead through a liquid democracy. The population of decision makers shifts from a small core at the center of a Legrand Star to a much wider network that is itself composed of smaller networks. The sports fans would cluster together, as would the anti-traffic constituency. Local business owners would petition their patrons to transfer their votes to them. If you followed the debate closely, you'd be free to vote directly on the stadium. But if you were busy with other priorities, you could pass your vote along to your friend who is obsessed with urban-planning issues. In fact, your friend might have a standing proxy from you for all urban-development votes—while another friend might represent you on any education initiatives, and another on fiscal reform.

The interesting thing about liquid democracies is that we already use this proxy strategy in our more casual lifestyle decisions. When you're trying to decide where to have dinner, you call up your foodie friend for advice, but there's another friend whose taste in music has never failed you, and yet another who is always coming up with great new novels to read. There's a natural division of labor that emerges in networks of friends or acquaintances; not all recommendations are created equal in the world of culture, because individual people vary in their tastes and expertise. When we make cultural decisions, we often offload those choices to the local experts in our network. Liquid democracies simply apply the same principle to political decisions.

Perhaps the most critical advantage of liquid democracies and participatory budgets lies in the way they expand the space of civic participation. The local experts who "club together" votes from their friends and neighbors occupy a new zone of engagement, somewhere between citizen and politician. They don't have to become full-time legislators for their particular expertise or passion to be useful to the community. Not surprisingly, this middle zone of participation resembles the effects of peer networks on the local news ecosystem. Proxy voters are like the influential bloggers and aggregators that have carved out a new space between media consumers and big news organizations. The net result is that the edges of government begin to blur: instead of the simple dichotomy of voter and elected official, a gradient of participation opens up, voter to proxy voter to general assembly member to elected official. Instead of choosing a candidate once every few years, ordinary citizens have a platform by which to participate in the decision-making process directly, according to their interests and their expertise. Yes, disagreements flourish when you widen the pool of participation. Cranks and crazies suddenly have a place at the table. But we know enough about these systems now to design social architectures that encourage and amplify the positive deviants, and discourage the negative ones. The officials want us to believe that they know best, and as individuals, they might just be right. But in the long run, diversity trumps ability.

Conscious Capitalism

In the fall of 2004, a Bentley College business professor named Rajendra Sisodia began researching a new project on the sorry state of contemporary marketing, working with two colleagues, David Wolfe and Jagdish N. Sheth. Sisodia had written several books that examined successful marketing campaigns, but this time around, his research had taken him in a new direction, to companies that had managed to thrive without much traditional marketing, like hiring celebrities to endorse their products, running Super Bowl ads, or putting up elaborate pavilions at trade shows. Despite this apparent lack of interest in marketing their brands, these companies had managed to attract fierce brand loyalty among their customers. The companies came from different sectors of the economy: grocery chains such as Whole Foods and Wegmans; athletic footwear manufacturer New Balance; Starbuck's. They were rarely the biggest firm in their industry, but they had something else that mattered as much as sheer size: a personal connection that consumers felt to their products and stores.

Sisodia and his colleagues began investigating these positive deviants to figure out what made them so successful at winning over the hearts (and wallets) of their customers. It turned out that the companies shared a set of core values that distinguished them from most of their rivals. For starters, unlike most corporations, they did not consider their ultimate responsibility to be "maximizing shareholder value." When management had to make key decisions, they didn't focus exclusively on how those decisions would play on Wall Street. Instead, the companies adhered to a "stakeholder" model, whereby decisions had to reflect the varied interests and needs of multiple groups: customers, employees, managers, shareholders, and even the communities that surrounded the company's stores or offices or factories. Most companies simply steered their operations to beat Wall Street expectations each quarter, thus ensuring a rising stock price, but these outlier firms greatly diversified their definitions of success. Good decisions were ones that satisfied all the stakeholders, not just the individuals who happened to own stock in the company.

Sisodia's positive deviants also broke from the deeply hierarchical structure that characterizes most modern corporations. They valued bottom-up decisions as much as those made in the executive suite. Whole Foods, for instance, gives its individual stores—and teams within each store—a great deal of autonomy in making decisions about how the store should be run, reflecting the local needs of the community around it, and the sensibilities of the employees themselves. Others created work environments where any employee of the company, no matter how low on the totem pole, can have direct access to the senior management. Flattening hierarchies might seem like a gesture of respect for those at the bottom of the chain, but the companies weren't just trying to create a friendlier work environment. They were deliberately trying to diversify the sources of information flowing through the management team, keeping them from losing touch with emerging problems, or missing out on ingenious new solutions that might arise from the factory floor or the checkout counter.

The flattened hierarchies of these companies carried over into their compensation packages as well. At Whole Foods, no one is allowed to make more than nineteen times the wages of the average worker. (For U.S. companies, the average ratio is more than four hundred to one.) But they don't just keep the ratios low; they also widen the number of decision makers involved in compensation. Google has long had a tradition of "peer bonuses," by which associates can spontaneously grant a colleague a $500 bonus for a cool idea, or a week of serious overtime crunching, or the completion of a long project. (The bonuses have to be approved by a supervisor, but the vast majority of them go through.) Each Whole Foods store is divided into teams—produce, checkout, meats—who are rewarded team bonuses based on performance, shared equally by all the members. But teams can opt to distribute some of their rewards to other teams in the store. Whole Foods calls this practice "gain-sharing." More than 30 percent of team bonuses are ultimately shared with colleagues on other teams.

Whole Foods also practices radical transparency where compensation is concerned. Every employee's pay package is visible to everyone else in the organization. Making that data widely available means any perceived inequities in compensation are quickly exposed and corrected, if need be. As Whole Foods CEO John Mackey writes, "Nothing de-motivates more quickly than the perception that the compensation system is unfair and rigged."

It was clear from Sisodia's research that these stakeholder-driven firms created better work environments for their employees than did traditional corporations. The question was how they fared on the bottom line. Sisodia and his colleagues decided to track the stock performance of the dozen or so public companies that best embodied these new values. These were all companies that had chosen a path that defied many core principles of modern capitalism: they paid their average employees much higher wages than their competitors did, and they resisted the pull of immense executive packages; they placed as much emphasis on the needs of their surrounding communities as they did on the needs of their shareholders. They democratized the decision-making process inside the organization.

Yet despite those strategies, the stakeholder-driven firms not only managed to stay in business, but actually outperformed the market by an extraordinary margin. In the ten years leading up to 2006, the public stakeholder firms had generated a 1,026 percent return for their investors, compared with the S&P 500's 122 percent return. By refusing to focus on maximizing shareholder value, they had created eight times more value for their shareholders.

John Mackey has come to call the overall philosophy shared by these firms "conscious capitalism." "If business leaders become more conscious of the fact that their business is not really a machine," he writes, "but part of a complex, interdependent, and evolving system with multiple constituencies, they will see that profit is one of the most important purposes of the business, but not the sole purpose. They will also begin to see that the best way to maximize long-term profits is to create value for the entire interdependent business system." In other words, empowering your employees and extending your stakeholder network is not just an honorable thing to do. It's ultimately a smart business decision as well.

Conscious capitalism is what happens when peer-progressive values are applied to corporate structures. Corporations are technically owned by their shareholders, who may meet once a year for procedural votes; every now and then a proxy fight breaks out over the overall direction or management of the company. But the overwhelming majority of the key decisions within the organization are made by CEOs and chairmen and senior management. The market that surrounds most corporations is a diverse web of competition and collaboration. But look inside the corporate walls and you'll find a Legrand Star. This has always been a contradiction at the heart of Hayek-influenced libertarianism; it prioritizes decentralized decision making over all other configurations except when it comes to the central organizational structure of capitalism itself: the corporation.

But if corporations begin to look more and more like peer networks, that contradiction fades away. Decentralized networks of collaboration compete with other decentralized networks in the larger network of the market itself. This is not socialism; there are still private firms and competition and price signaling and all the other wonders of capitalism that Hayek rightly celebrated. The difference is that the firms themselves are not run like fiefdoms or politburos.

The beauty of the peer-progressive approach to corporate organization is that it addresses many of the prevailing critiques of modern capitalism. If decreasing trust in government is one disturbing trend of the past two decades, equally disturbing is the inequality gap that has opened up over that same period in the private sector. Countless studies have shown that runaway executive pay results from the small, incestuous circle of corporate boards, largely made up of other overpaid executives at other corporations. The decision of how best to distribute a firm's profits is almost never made by the shareholders, much less the employees. The decision, instead, comes from above. Allowing employees and shareholders to participate in major decisions—particularly ones involving how the firm shares its profits—could go a long way in reducing the inequality gap. And by empowering local decision making among employees, conscious capitalism creates a culture in which individual stores can reflect the idiosyncratic needs of their communities.

Americans have understandably mixed feelings about many sectors of the current economy: manufacturing jobs seem to be on an irreversible, secular decline, though the automobile industry appears to be having a renaissance of late; the service industry is growing, but not as fast as some would like; media and advertising have their growth stories and their death spirals, depending where you look. Yet the region of the U.S. economy that everyone agrees is the envy of the world remains the tech sector, led by the giants of Silicon Valley. The question is, Why? Since the early 1970s, U.S. share of both patents and Ph.D.s has been in steady decline. (In 1970, more than 50 percent of the world's graduate degrees in science and engineering were issued by American universities.) Since the mid-1980s, a long progression of doomsayers have warned that our declining market share in the patents-and-Ph.D.s business augured dark times for American innovation. The specific threats have changed: it was the Japanese who would destroy us in the eighties; now it's China and India.

But what actually happened to American innovation during that period? Instead of slow decline, the tech industry skyrocketed: Apple, Netscape, Amazon, Google, Blogger, Wikipedia, Craigslist, TiVo, Netflix, eBay, the iPod, iPhone, iPad, Xbox, Facebook, Twitter. If you measure global innovation in terms of actual lifestyle-changing hit products, and not just grad students, the United States—and Silicon Valley in particular—has been lapping the field for the past twenty years.

There are many potential explanations for that success, including the unique social chemistry of the Bay Area, with its strange cocktail of engineering geeks, world-class universities, and countercultural experimentation. But the organizational structure of most Silicon Valley firms also deserves a great deal of the credit. Early tech companies in the region—starting with Fairchild Semiconductor and its mammoth offspring, Intel—deliberately broke with the hierarchical social architecture of East Coast firms: in executive privileges, decision making, and compensation. Stock options, for instance, were distributed sparingly for most of the twentieth century, until the more egalitarian operations in the Bay Area began using them as a standard part of an employee's compensation package. The options structure meant that the entire organization had a vested interest in the company's success—and when things went well, a larger slice of the Silicon Valley population was free to experiment with new start-up ideas thanks to the payday from the first IPO.

In the early 1980s, *Esquire* magazine sent Tom Wolfe out to Silicon Valley to profile Intel founder Robert Noyce and document the emerging culture of the region. The shift from hierarchy to peer network was visible even then, in the days before computer networks became essential to the region's businesses:

Corporations in the East adopted a feudal approach to organization, without even being aware of it. There were kings and lords, and there were vassals, soldiers, yeomen, and serfs, with layers of protocol and per-quisites, such as the car and driver, to symbolize su-periority and establish the boundary lines . . . Noyce realized how much he detested the eastern corporate system of class and status with its endless grada-tions, topped off by the CEOs and vice-presidents who conducted their daily lives as if they were a cor-porate court and aristocracy. He rejected the idea of a social hierarchy at Fairchild. Not only would there be no limousines and chauffeurs, there would not even be any reserved parking places. Work began at eight A.M. for one and all, and it would be first come, first served in the parking lot for Noyce, Gordon Moore, Jean Hoerni, and everybody else. "If you come late," Noyce liked to say, "you just have to park in the back forty." And there would be no baronial office suites. . . . [Later] at Intel, Noyce decided to eliminate the notion of levels of management altogether.

Most Silicon Valley companies, like their brethren in the conscious-capitalism movement, have traditional owner-ship structures, even if they buck tradition in countless other ways. But a more radical version of the peer-network approach to capitalism exists in the form of employee-owned businesses, or EOBs. These organizations are true creatures of capitalism, not some kind of socialist collective or a nonprofit NGO. They want to make money as much as the most ruthless Wall Street firm. The difference is that the shareholders are for the most part the employees themselves. Most companies today are run by management, and owned by some combination of management and outside shareholders who have almost no participation in the day-to-day operation of the firm. An EOB brings that shareholder base inside the organization. When the time comes for important shareholder votes on the future of the firm—electing the board, for instance—it's the employees who make the decision, not some pension fund or day trader somewhere on the other side of the planet.

Two years ago, three researchers at the Cass Business School completed one of the first comprehensive studies of how EOBs stack up against traditional corporate structures. They looked at financial data from about fifty EOBs in the United Kingdom and compared it with roughly two hundred traditional firms there. The period they studied spanned from 2005 to 2008, which allowed them to analyze each firm's performance during periods of both macro-economic growth and decline. They found that small and midsized EOBs had a distinct advantage over traditional firms on a number of fronts. EOBs generated more jobs, were more productive per employee, and were far more resilient than traditional corporations during the economic downturn. The EOBs were just as profitable as their traditional equivalents, but more egalitarian in their compensation. Tellingly, the Cass study found that EOBs performed even better if the employees were actively engaged in "decisions that influence governance, strategy-making and operations." The more participatory, the more profitable.

The Cass study inspired a new set of initiatives announced by Deputy Prime Minister Nicholas Clegg in early 2012, to encourage the growth of EOBs in the United Kingdom and reduce some of the legal and tax barriers that can make them more challenging to form and maintain. "What many people don't realize about employee ownership," Clegg remarked, "is that it is a hugely underestimated tool in unlocking growth."

Why do the benefits of employee ownership become less pronounced as the company grows in size? We don't have a definitive account yet, but one plausible explanation would be the information funnel that has traditionally restrained direct democracies. A multinational corporation with billions of dollars in revenue and offices around the world presents information-overload challenges that a small firm doesn't. Employees in a small firm can contribute productively to overall decisions because the firm itself is a knowable problem at that scale. On a multinational scale, however, it gets much harder for a mid-level manager in the Toledo office to have a nuanced understanding of what the company's strategies should be for subcontracting Chinese manufacturing talent. But this, of course, is precisely the problem that peer networks were designed to solve. If key decisions were shaped by proxy votes, liquid democracy–style, whereby clusters of expertise and influence could be weighted in the overall vote tally, then some of the scale problems faced by EOBs might be overcome.

What goes for private-sector organizations goes for the public sector as well. Consider the question of performance-based pay in the public school system. For years, reformers have argued that teachers who succeed in the classroom should see their compensation increase. It seems like a simple enough premise, one that has worked fairly well for capitalism over the years: linking raises to on-the-job performance. No one has ever agitated for performance-based pay in the private sector, because doing so would be akin to demanding that Starbucks sell caffeinated beverages; performance-based pay is part of the fundamental ethos of capitalism. Yet the teachers' unions have long resisted the idea, which has led many a critic to complain about their socialist values, or to suggest that they are more interested in protecting their less talented colleagues than they are in educating kids.

It turns out the unions have a more nuanced objection to performance-based pay, one that revolves around the camaraderie and collaboration that one so often finds in great schools. Performance-based compensation doesn't just encourage you to teach to the test; it also encourages you to keep your pedagogical tactics to yourself. If you've figured out a brilliant way to convey the subtle logic of long division to fourth-graders, you're literally incentivized to withhold that technique from your fellow fourth-grade teachers. If you're trying to land in the top percentile to ensure your $5,000 raise, you've got every reason to keep your competitors in the dark about your new strategies in the classroom.

But imagine that school is structured more like an employee-owned business than a traditional corporation, or like the gainsharing teams at Whole Foods. Teachers would be shareholders in the school, not employees. Performance would be rewarded—not through individual raises, but through an increase in the overall budget of the school, which would then be passed down to all the teachers through bigger paychecks. Just as we saw with Race to the

Top, that reward would be fronted by the state, but it would play the same role that increases in publicly held shares do in the private sector. The rising tide of shareholder value would lift all boats. That kind of incentive structure would encourage better teaching and better collaboration with other teachers. The school would be a peer network at its finest: a group of minds gathered together to tackle an important problem, where promising ideas were both rewarded and free to circulate through the network.

Note again where the peer progressive comes down on this issue. The teachers' unions are wrong to resist performance-based compensation, yet they're right that raw competition in their ranks is toxic, given the goals of public education. The libertarians, on their end, are wrong that public schools should be turned into mini-corporations, driven exclusively by self-interest. The peer progressive doesn't want to turn our education system over to the private markets, because private markets are not omnipotent. But peer progressives want to do away with the bureaucracies as well as the union mentality. They want schools to be run like EOBs, where teachers are shareholders in an enterprise that grows more valuable as it reaches its goal. Yes, it is not raw capitalism if the state pays out the dividends, but it is just as far from bureaucratic socialism as well. The state plays a key role, but that role is limited to establishing rewards and incentives that encourage better—and more collaborative—teaching.

In February 2012, Facebook filed its S-1 with the Securities and Exchange Commission, justifying and describing its plans for an initial public offering. The document included a revealing letter from Facebook cofounder Mark Zuckerberg, outlining the company's core mission and warning potential shareholders that the company would prioritize that long-term mission over short-term opportunities to increase the share price.

The Facebook mission can be boiled down to the old E. M. Forster slogan: "Only connect." The company wants to strengthen the social ties that allow humans around the planet to connect, organize, converse, and share. At one point, Zuckerberg writes:

> By helping people form these connections, we hope to rewire the way people spread and consume information. We think the world's information infrastructure should resemble the social graph—a network built from the bottom up or peer-to-peer, rather than the monolithic, top-down structure that has existed to date.

In other words, the Facebook platform is a continuation of the Web and Internet platforms that lie beneath it: it is a Baran Web, not a Legrand Star. And it considers the cultivation and proliferation of Baran Webs to be its defining mission. Zuckerberg clearly believes that the peer-network structure can and should take hold in countless industries, in both the private and public sectors. As he remarked on another occasion: "Over the next five years, most industries are going to get rethought to be social, and designed around people."

The peer progressive encounters the Facebook mission with mixed feelings. On the one hand, the conviction that peer networks can be a transformative force for good in the world is perhaps the core belief of the peer-progressive worldview. So when you hear one of the richest and most influential young men in the world delivering that sermon—in an S-1 filing, no less—it's hard to hold back from shouting out a few hallelujahs. But there's a difference here, one that makes all the difference. The platforms of the Web and the Internet were pure peer networks, owned by everyone. Facebook is a private corporation; the social graph that Zuckerberg celebrates is a proprietary technology, an asset owned by the shareholders of Facebook itself. And as far as corporations go, Facebook is astonishingly top-heavy: the S-1 revealed that Zuckerberg personally controls 57 percent of Facebook's voting stock, giving him control over the company's destiny that far exceeds anything Bill Gates or Steve Jobs ever had. The cognitive dissonance could drown out a Sonic Youth concert: Facebook believes in peer-to-peer networks for the world, but within its own walls, the company prefers top-down control centralized in a charismatic leader.

If Facebook is any indication, it would seem that top-down control is a habit that will be hard to shake. From Henry Ford to Jack Welch to Steve Jobs to Zuckerberg himself, we have long associated corporate success with visionary and inspiring executives. But the empirical track record of the conscious capitalists and employee-owned businesses suggests that we might have been focusing on the wrong elements all along. Yes, leaders can endow a company with vision and purpose. But corporations that funnel too much power into the hands of their executive team confront the exact same problems that Hayek saw among the central planners of socialism. Like communities and governments and software platforms, corporations perform better when they expand and diversify the network of participants involved in setting goals, generating ideas, and making decisions. Capitalism helped us see the value of decentralized networks through the price signaling of markets. The next phase is for capitalism to apply those lessons to the social architecture of corporations themselves.

Conclusion

THE PEER SOCIETY

The U.S. House of Representatives may indeed be a wasteland of partisan hostility, but you wouldn't have known it looking at the lineup of cosponsors who put their name behind a new piece of legislation, formally known as H.R. 3261, in October 2011. The bill was initially backed by nine Republicans and four Democrats; within two months, it had accumulated thirty-one cosponsors, with an almost equal distribution of support between the two parties, including representatives from all around the country: red and blue states, urban and rural districts. What common cause managed to break down the walls of party polarization? Support for the troops, perhaps? A plan to make the United States more energy independent? No, the cross-party rallying cry turned out to be the threat of Internet piracy.

H.R. 3261 was the Stop Online Piracy Act; the bill proposed to radically strengthen the government's ability to punish—or in some cases shut down—websites that promoted unauthorized copyrighted material. Under the act, streaming of movies or music without approval of the copyright owners would become a felony, punishable by up to five years in prison. Courts would be empowered to demand that search engines remove all links to offending sites. The Department of Justice would be allowed to shut down an entire top-level domain because of the activity of a single rogue user.

Wherever you came down on the question of online piracy, one thing was clear: SOPA marked the single largest attack to date on the core principles of the Baran Web, at least on U.S. shores. It was a bold attempt to restore top-down control on a system that had been designed to resist those forces. SOPA was the unholy alliance of two Legrand Stars—the federal government and the largest media corporations on the planet—combining forces to rein in a peer network that had fundamentally undermined their authority. For several months after the bill was first proposed, it looked to be a sure winner. The Senate version sailed through the Judiciary Committee with a unanimous voice vote; by the end of 2011, almost half the Senate had signed on as a cosponsor of the bill.

But then something peculiar happened. In the language of the Terminator movies, the network became self-aware. It started to fight back.

In the late fall, posts and comment threads at influential sites—Reddit and BoingBoing, along with the New York–based venture capital firm Union Square Ventures—began describing the bill's authoritarian provisions. A satirical site and mock movement named freebieber.org noted that teen heartthrob Justin Bieber, who began his career by uploading YouTube clips of himself singing unauthorized covers of copyrighted songs, would be considered a felon under the law. Legal scholars, Net theorists, entrepreneurs, and artists penned multiple editorials warning of the chilling effect that the bill would have on all user-generated websites. The growing backlash culminated in a historic day of online protest in January 2012, when thousands of sites went offline, or replaced their front doors with anti-censorship messages—most notably, Wikipedia, which ran on its opening page the Orwellian imperative: "Imagine a world without access to free information." Some estimate that more than a billion people saw anti-SOPA messages during the day. By the end of the week, the bill was dead in the water.

It goes without saying that the SOPA rebellion made manifest the power of peer networks as a form of Digital Age activism; in a few months, without a single dominant organization or official leadership, the movement managed to derail a major piece of legislation, building awareness for the cause on a global scale. The bill's supporters griped that the protests had been spearheaded by rapacious Silicon Valley corporations who valued their own profits over the rights of artists to make a living, but in truth the Googles and Facebooks of the world were relative latecomers to the movement. And of course the most visible form of protest— the Wikipedia blackout—took place on a nonprofit site whose vast audience exists entirely because of the free labor supplied by its contributors.

But the SOPA protest made something else visible, something that was particularly vivid in the days after the blackout, as politicians began running away from the bill they had originally supported. Normally, when a major piece of legislation either passes or dies, it's relatively trivial for the media commentators to "score" the event as a victory for either the Left or the Right. (Horse-race political journalism makes this kind of assessment the default mode of most punditry.) But the collapse of SOPA support was just as nonpartisan as its original sponsorship. Even before the blackout, both Rand Paul and Nancy Pelosi had voiced opposition to the bill. As the protests began mounting in January, the White House announced that it would not support the bill in its current form, because of the threat it posed to the Internet's open architecture. After the blackout, the Right tried to paint the bill as another case of the Democrats' close ties to Hollywood, thanks in part to former Democratic senator Chris Dodd, who, as the head of the Motion Picture Association of America, had been the most vocal supporter of the bill. But in the end, it was impossible to paint the SOPA failure as a victory for either the Democrats or the Republicans, because the values at stake were not fundamentally aligned with either party.

SOPA failed because a growing number of people in the world have come to recognize the power and utility of the Internet's decentralized, peer-to-peer architecture. That's what they were fighting to defend: not corporate interests or music piracy, but rather the social architecture that has made the Internet such an extraordinary platform for human connection. Yet believing in the power of peer networks is not a core value of either political party in the United States. That's why it was so easy for Congress to build a bipartisan consensus in favor of SOPA, and why the ultimate backlash against the bill was so hard to "score" along traditional party lines. There is a growing constituency of people who realize that something wonderful has happened to the world, thanks to the peer networks of the Internet and the Web, people who want to see more of those networks, not less. The problem is—in the United States, at least—that constituency doesn't have a natural home in either party. Judged by the definitions of the existing political establishment, the SOPA protests seemed to come from nowhere. But if you widen the definitions beyond the two-party polarity, the protests made perfect sense: they were the first great awakening of the peer-progressive movement.

Institutions are the giant redwoods of civic life. They out-live everything around them: individuals, fashions, corpo-rations, technologies. The United States is still living with an operating system that was conceived and designed be-fore railroads were invented. This longevity is, of course, one source of the institution's strength. It has lasted that long, so surely it deserves some measure of respect. But that longevity also means that the institution is often locked into ways of operating that cannot easily adapt to changes in the external world. Many of the problems that we con-front today come from the social disconnect of institutions that have outlived their original relevance.

The two-party system in American politics has been in place for more than two hundred years, with a few brief pe-riods when third or fourth parties were able to attract a meaningful audience. Over time, of course, the values associated with each party have shifted. The last great realignment took form in the late 1960s, when white southerners left the Democratic Party in droves after the civil rights reforms and Goldwater's failed campaign ignited the more libertarian wing of the party, eventually driving away the more moderate Rockefeller Republicans. Because many of us have lived our entire adult lives with these two political institutions, we naturally assume that between them they must cover a significant span of the political spectrum.

But think about the specific values that we have seen associated with the peer-progressive worldview. Peer progressives are wary of excessive top-down government control and bureaucracy; they want more civic participation and accountability in public-sector issues that affect their communities. They want more choice and experimentation in public schools; they think, on the whole, that the teachers' unions have been a hindrance to educational innovation. They think markets can be a great force for innovation and rising standards of living, but they also think that corporations are far too powerful and top-heavy in their social architecture. They believe the rising wealth and income gaps need to be restored to levels closer to those of the 1950s. They believe that the campaign-finance system is poisoning democracy, but want to retain an individual's right to support candidates directly. They want lower prices for prescription drugs without threatening the innovation engine of the pharmaceutical industry. They are socially libertarian, and consider diversity to be a key cultural value. They believe the decentralized, peer-to-peer architecture of the Internet has been a force for good, and that governments (or corporations) shouldn't mess with it.

If you look at these values through the lens of either party, the lines blur: the values read like a mishmash of positions drawn seemingly at random from either side of the political aisle. But through the lens of peer progressivism, they all come into focus, because the values flow directly from a core set of beliefs about the power and effectiveness of peer networks, in both the private and public sectors. The number of individuals and groups that are actively building new peer-progressive organizations is still small, but the values associated with the movement are shared much more widely throughout the population. Yet because the parties are institutions stuck in older ways of organizing the world, the electorate has to distort the square peg of its true political worldview to fit the round holes of the two parties. (This is, not coincidentally, exactly the way new scientific paradigms come about: anomalous data start to appear, behavior that doesn't fit existing expectations.) The mismatch between core values and party organization is surely one reason why dissatisfaction with the existing political leadership—particularly in the hyperpartisan House and Senate—is at an all-time high. The parties have failed to adapt to emerging attitudes and beliefs within their constituencies.

It is a measure of how perplexing the peer-progressive worldview is to the current taxonomies that the national political figures who have done the most to advance peer-progressive experiments or ideas are probably Bernie Sanders, Rand Paul, and certain segments of the Obama administration, particularly in the Department of Education and in Obama's technology team. By conventional political accounts, Sanders, Paul, and Obama exist on completely different spots in the spectrum: socialist independent, libertarian Republican, and center-left Democrat. Yet somehow, despite their professed allegiances, these three have shown a willingness to buck the orthodoxies of their parties and remain receptive to new ideas.

And that is ultimately what being a peer progressive is all about: the belief that new institutions and new social architectures are now available to us in a way that would have been unthinkable a few decades ago, and that our continued progress as a society will come from our adopting those institutions in as many facets of modern life as possible. No doubt there will be places where the approach turns out to be less effective. It may well be that certain pressing problems—climate change, military defense—require older approaches or institutions. The peer-progressive framework is in its infancy, after all. We don't yet know its limits.

Many of the most promising peer networks today utilize advanced technology, but from a certain angle, they can be seen as a return to a much older tradition. The social architectures of the paleolithic era—the human mind's formative years—were much closer to peer networks than they were to states or corporations. As E. O. Wilson writes in *The Social Conquest of Earth*, "Hunter-gatherer bands and small agricultural villages are by and large egalitarian. Leadership status is granted individuals on the basis of intelligence and bravery, and through their aging and death it is passed to others, whether close kin or not. Important decisions in egalitarian societies are made during communal feasts, festivals, and religious celebrations. Such is the practice of the few surviving hunter-gatherer bands, scattered in remote areas, mostly in South America, Africa, and Australia, and closest in organization to those prevailing over thousands of years prior to the Neolithic era." Defenders of the free market have long stressed the "natural" order of competition, drawing on a loose interpretation of Darwin's "survival of the fittest." But as Darwin himself understood, webs of collaboration and open exchange have always been central to evolutionary progress, never more so than in the history of our intensely social species.

The fact that a corporation (or a government bureaucracy) would have been bewildering to our paleolithic ancestors does not in itself mean those modern inventions are worthless. (After all, the Internet would have perplexed them just as much.) But it may help explain why so many people have been drawn to participate in peer networks, despite the lack of traditional monetary rewards. There is something in the collaborative, egalitarian structure of these systems that resonates with the human mind, an echo of our deep history as a species.

The new technological regime of the Internet and its descendants—what Manuel Castells astutely described as the "network society" many years ago—has finally given us an opportunity to put peer networks to work in a modern context. That is partly because the new technology allows us to build things that were technically impossible even a few decades before. But the new technology has also helped us understand how these peer networks can work, and given us the confidence to have faith in systems that would have seemed like utopian fantasies before the practical success of Wikipedia or Kickstarter. The technology makes it easier for us to dream up radical new approaches—crowdfunded art, liquid democracies—and it makes it easier for us to build them.

The modern regime of big corporations and big governments has existed for the past few centuries in an artificial state that neglected alternative channels through which information could flow and decisions could be made. Because we were locked into a Legrand Star mind-set, we didn't build our businesses and our states around peer networks that could connect us to a much more diverse and decentralized group of collaborators. Instead, we created a mass society defined by passive consumption, vast hierarchies, and the straight lines of state legibility. It didn't seem artificial to us, because we couldn't imagine an alternative. But now we can.

On November 11, 2001, a programmer and engineer named Greg Lindahl uploaded an initial entry defining peer-to-peer software for a new collaborative encyclopedia, named Wikipedia, that had just launched a few months before.

> A peer-to-peer application architecture is one in which each node, or logical entity, in the global application (consisting of all of the interoperating peer nodes) may act as either client or server, whether in

turn (serial roles) or simultaneously (parallel roles), [and] may be termed a peer-to-peer application architecture, and such applications are often known as peer-to-peer, or P2P "apps."

It was a terse, technical entry, with somewhat questionable grammar; for a non-programmer, it was probably useless as a definition. But it was a start. A few months after Lindahl had created his entry, another Wikipedia contributor posted a structural question about how the entry should be organized. The next day, another contributor added a link to a popular file-sharing site that embodied peer-to-peer values. Five months passed before another contributor edited the document, this time just correcting a spelling mistake. But then the waves of alterations start to roll in at an accelerating pace. A user identified as "DK" penned an extensive rewrite in mid-August 2002, and before long multiple additions or comments were arriving on the entry each month.

Ten years later, the peer-to-peer entry on Wikipedia is almost five thousand words long, twenty times longer than the entry on the Encyclopædia Britannica website. Over that decade, 1,694 authors contributed to this single entry, making a total of 3,204 revisions. Some of those revisions turned out to be misguided; some furthered an agenda that other users found problematic; others were sloppy with their facts or their spelling. Lindahl himself ultimately stopped contributing to Wikipedia, driven off by what he called the "deletionistas." But over time, the revisions tightened the prose and expanded the scope of the entry. The end result is a rich, informative, well-structured document that covers the technical elements of peer-to-peer networks, but also includes sections on their social and economic impact, and the historical context of their invention. Near the end of the entry, the text even suggests that peer-to-peer networks are increasingly being used to describe human-to-human interactions, including notions of peer governance and peer production.

The media like to highlight stories of Wikipedia abuse: the scurrilous attack that a user has added to a rival's biographical entry; the endless fighting over the content of the abortion entry, or the entry on the Iraq War. But these stories are like the plane crashes of the information world: the sensational news that distracts us from the steady, incremental miracle that works astonishingly well almost all the time. Wikipedia is a living book, growing smarter and more comprehensive every day, thanks to the loosely coordinated actions of millions of human beings across the planet. Like Hayek's marketplace, it works as well as it does precisely because no single individual or group understands the whole of it. But unlike Hayek's marketplace, it has no price signaling, or traditional rewards, beyond the satisfaction of being able to add a small bit to the sum total of human knowledge. We have Wikipedia because the Internet and the Web made it easy and cheap to share information, and because they allowed people to experiment with new models of collaboration while minimizing the risks of failure. It took a peer network to solve the problem of how to come up with the best description of peer-to-peer software, just as a peer network solved the higher-order problem of building a global encyclopedia. Only a decade ago, as Greg Lindahl was writing out his initial definition of P2P, the success of those solutions would have seemed like a fantasy, the sort of thing you might expect from the most Panglossian of sci-fi writers. Today, that living book is a reality.

To be a peer progressive, then, is to live with the conviction that Wikipedia is just the beginning, that we can learn from its success to build new systems that solve problems in education, governance, health, local communities, and countless other regions of human experience. This is why we are optimistic: because we know it can be done. We know a whole world of pressing social problems can be improved by peer networks, digital or analog, local or global, animated by those core values of participation, equality, and diversity. That is a future worth looking forward to. Now is the time to invent it.

Acknowledgments

Appropriately enough, this book is itself the product of a peer network of support and collaboration that extends back more than a decade. Shortly after my 2001 book *Emergence* was published, my friend Joi Ito posted a few open-ended musings online about what the decentralized networks of emergence might mean for the future of politics and democracy—sparking a conversation that ultimately produced a book of essays called *Extreme Democracy*. At the same time, my friend Kurt Andersen remarked that he wished *Emergence* had explored the social and economic implications of the theories I was describing. A few years later, in a casual e-mail exchange, the political scientist Henry Farrell encouraged me to write a book documenting the new politics that seemed to be coalescing around the online world. It took me a while, but eventually all these nudges came together to form *Future Perfect*. My thinking on these issues has been greatly expanded—if not downright borrowed—from conversations with Beth Noveck, Yochai Benkler, Fred

Wilson, Brad Burnham, Larry Lessig, Denise Caruso, John Mackey, John Geraci, Paul Miller, Roo Rogers, Rachel Botsman, Reid Hoffman, Perry Chen, Yancey Strickler, Clay Shirky, Stewart Brand, Howard Rheingold, Kevin Kelly, Jon Schnur, Raj Sisodia, Gordon Wheeler, Nick Grossman, Jay Haynes, Eric Liftin, John Battelle, and my mother, Bev Johnson. Special thanks to the group who were generous enough to comment on the manuscript in draft: Bill Wasik, David Sloan Wilson, Dan Hill, Henry Farrell, and my father and longtime political sparring partner, Stan Johnson.

As usual, my wife, Alexa Robinson, shared her invaluable talent for improving my sentences *and* my arguments. Whatever errors she—and the rest of the crew—managed to miss are my responsibility alone.

I am grateful to people at several institutions who have asked me to speak on these topics over the past few years: Princeton University, South by Southwest, Personal Democracy Forum, the Columbia University Graduate School of Journalism, the Royal Society of Arts, the Government 2.0 conference, and the Esalen Center for Theory & Research. I'm grateful as well to my editors at *Wired*, *Time*, *Prospect*, and *The Wall Street Journal* for allowing me to explore some of this material in print.

Thanks, as well, to the team at Riverhead, especially to my always astute editor Geoffrey Kloske, Laura Perciasepe, and Matthew Venzon. My wonderful lecture agents at the Leigh Bureau kept me busy during the writing of this book, but also led me to many new connections that enhanced the argument. And my literary agent, Lydia Wills, once again helped this idea find the right home, and encouraged me to set off in the new directions that I found myself following as I began writing.

This is my eighth book, but the first one not written in New York. Thanks to all our California friends, new and old, for creating such a wonderful environment to write and think over the past year.

May 2012
Marin County, California

Notes

INTRODUCTION. PROGRESS, ACTUALLY

The article on airline safety, "Airlines Go Two Years with No Fatalities," appeared in the January 12, 2009, edition of *USA Today*. My original post on air safety and subsequent coverage of the US Airways crash ran on the website BoingBoing; the first post can be found at http://boingboing .net/2009/01/14/for-once-news-about.html. William Langewiesche's *Fly by Wire: The Geese, the Glide, the Miracle on the Hudson* gives a thorough account of the Airbus 320 design and its role in the Hudson landing. Peter Thiel's "The End of the Future" appeared in the October 3, 2011, issue of *National Review*.

High school dropout rates and college enrollment: Between 1988 and 2008, the high school dropout rate for the United States declined from 14.6 to 9.3. (Source: "Trends in High School Dropout and Completion Rates in the United States: 1972–2008"; http://nces.ed.gov/pubs2011/20110 12.pdf.) During that same period, college enrollment increased from 30.3 to 39.6 percent. (Source: Pew Research Center.)

SAT scores: Many educational metrics in the United States appear to show the country flatlining over the past two decades, but the appearances are deceiving. Mean SAT scores of college-bound seniors, for instance, stood at 500 and 501 for critical reading and mathematics, respectively, in 1990; by 2010, they had inched ahead to 501 and 516. And yet, during that period, almost every single ethnic group outperformed the mean. For instance, Puerto Ricans jumped from 436 to 454 on the critical reading tests, while Asians jumped by almost 50 points on the mathematics exam. How can the average perform worse than the individual groups that together make up the average? The answer lies in a curious statistics phenomenon known as Simpson's paradox: Because the overall demographic composition of the American population has changed over the past twenty years—thanks in large part to immigration—the pool of SAT-takers now includes more groups that score, on average, lower than other groups. That shift weighs down the overall average, even though each individual group has actually been making steady progress during the period. The effect is even more pronounced on the National Assessment of Education Progress (NAEP) tests, which offer a broader survey of the U.S. population. For example, fourth-grade reading scores improved modestly between 1990 and 2010, moving from 210 to 220. Yet the sole demographic group that improved by only 10 points during that period were white students; black, Hispanic, and "other" ethnicities all improved by more than 20 points. (Source: College Board; NAEP.)

Juvenile crime: In 1988, the arrest rate for all juvenile crime in the United States stood at 8 percent; by 2008 it was less than 6 percent. The patterns for violent crime are similar, though in each case, the dramatic decline arrived in the past decade, after a severe spike in the late 1980s and early 1990s. (Source: Office of Juvenile Justice and Delinquency Prevention.)

Drunk driving: From 1991 to 2010, the rate of alcohol-impaired driving fatalities in the United States dropped by 48 percent. For drivers under twenty-one, the rate dropped by 58 percent. (Source: The Century Council.)

Traffic deaths: A total of 33,963 people died in car accidents in 2009, the lowest absolute number since 1954. Measured by total miles traveled, the fatality rate decreased from 1.73 to 1.14 between 1994 and 2009. (Source: National Highway Traffic Safety Administration.)

Infant mortality: In 1988, 9.4 infants died per 1,000 born in the United States; twenty years later, the rate had been reduced to 6.7 per 1,000. (Source: UN Inter-agency Group for Child Mortality Estimation, childmortality.org.)

Life expectancy: In 1990, the average American could expect to live to age seventy-five; by 2010, life expectancy had increased to seventy-eight.

Per capita crude oil consumption: Americans consumed approximately 24 barrels of crude oil per capita in 1988; in 2008, the number was 21. Per capita crude consumption peaked in the early 1970s, at almost 30 barrels. (Source: Energy Information Administration, Annual Energy Review 2008, http://www.eia.gov/FTPROOT/multifuel/038408.pdf.)

Workplace injuries: In 1992, 6,217 Americans were killed on the job; in 2010, despite an increase in the overall population during that period, the number was 4,690. (Source: U.S. Bureau of Labor Statistics.)

Air pollution: Between 1990 and 2010, carbon monoxide in the air declined by 73 percent; ozone by 17 percent; lead by 83 percent; and nitrogen dioxide by 45 percent. (Source: United States Environmental Protection Agency.)

Divorce: Between 1990 and 2008, the divorce rate per 1,000 married women fell from 20.9 to 16.9, after peaking at 22.6 in 1980. (Source: The National Marriage Project.)

Male-female wage equality: In 1990, the average American woman with a full-time job earned 70 percent of what an average male earned; by 2009, she earned 80 percent of her male counterpart's salary. (Source: U.S. Bureau of Labor Statistics.)

Charitable giving: Even after the recession of 2008–2009, annual U.S. charitable giving stood at $303 billion, up from approximately $170 billion in 1990, in inflation-adjusted dollars. (Source: Giving USA.)

Voter turnout: Both the 2008 and 2004 federal elections had the highest voter turnout (56.8 percent and 55.3 percent, respectively) since the 1968 elections (60.8 percent). (Source: Federal Election Commission.)

Per capita GDP: In 2008, per capita GDP in the United States stood at almost $38,000; in 1988, it was $28,000. (Source: World Bank.)

Teen pregnancy: Between 1986 and 2006, the fertility rate per 1,000 U.S. teenagers dropped from 50 to 43. (Source: World Bank.)

The two essential books on our strange unwillingness to accept the progressive trends around us are Gregg Easterbrook's *The Progress Paradox: How Life Gets Better While People Feel Worse* and Matt Ridley's *The Rational Optimist: How Prosperity Evolves*. On long-term trends in human violence, see Steven Pinker's *The Better Angels of Our Nature: Why Violence Has Declined*.

I. THE PEER PROGRESSIVES

For more on the history of the Legrand Star, see "The Longest Run: Public Engineers and Planning in France," by Cecil O. Smith, Jr., published in *The American Historical Review*. The "legible" vision of state hierarchy is powerfully analyzed in James C. Scott's *Seeing Like a State: How Certain Schemes to Improve the Human Condition Have Failed*. For a fascinating discussion of Scott's theories and their relationship to Friedrich Hayek, see "Forests, Trees, and Intellectual Roots," by J. Bradford DeLong, at http://econ161.berkeley.edu/econ_articles/reviews/seeing_like_a_state.html, along with Henry Farrell's response, "Seeing 'Seeing Like a State'" at http://crookedtimber.org/2008/02/05/seeing-like-seeing-like-a-state/.

Hayek's arguments are nicely summarized in his 1945 essay "The Use of Knowledge in Society," which appeared in *The American Economic Review*. Jane Jacobs's attack on centralized planning appears in her classic work, *The Death and Life of Great American Cities*.

For a comprehensive history of the birth of the Internet, see *Where Wizards Stay Up Late*, by Matthew Lyon and Katie Hafner, as well as Stewart Brand's interview with Paul Baran, "Founding Father," in *Wired*. I first came across the concept of "positive deviance" in the article "Design Thinking for Social Innovation," by Tim Brown and Jocelyn Wyatt, in the *Stanford Social Innovation Review*. For more on the approach, see the website of the Positive Deviance Initiative at http://www.positivedeviance .org/. For more on Marian Zeitlin's original work, see *Positive Deviance in Child Nutrition (with Emphasis on Psychosocial and Behavioural Aspects and Implications for Development)*, coauthored with Hossein Ghassemi and Mohamed Mansour. The key books and essays that have shaped my thinking on the power of peer networks and the framework of peer-progressive values include Yochai Benkler's *The Wealth of Networks*; Beth Noveck's *WikiGovernment*; Carne Ross's *The Leaderless Revolution: How Ordinary People Will Take Power and Change Politics in the 21st Century*; Howard Rheingold's *Smart Mobs*; Clay Shirky's *Here Comes Everybody: The Power of Organizing Without Organizations*; Ori Brafman and Rod Beckstrom's *The Starfish and the Spider: The Unstoppable Power of Leaderless Organizations*; Tim O'Reilly's "The Architecture of Participation"; Henry Farrell and Cosma Shalizi's "Cognitive Democracy"; and just about everything written by Manuel Castells, starting with *The Rise of the Network Society*. Many of these themes are explored in my books *Emergence* and *Where Good Ideas Come From*. For more on the gift economy, see Kevin Kelly's essay "The Web Runs on Love, Not Greed," in the January 4, 2002, edition of *The Wall Street Journal*. The figure of some $1.5 billion passing through crowdfunding sites in 2011 is from *Forbes*, http://www.forbes.com/sites /suwcharmananderson/2012/05/11/crowdfunding-raised-1-5bn-in-2011-set-to-double-in-2012/.

II. PEER NETWORKS AT WORK

Communities. The Maple Syrup Event

For more on 311 and other urban technology platforms, see my essay "What a Hundred Million Calls to 311 Reveal About New York," in the November 2010 issue of *Wired*, and Vanessa Quirk's essay "Can You Crowd-source a City?" (http://www.archdaily.com/233194/can-you-crowdsource -a-city/). A number of sites and mobile apps are innovating in the area of peer-network urbanism; an overview of many of them is available at the DIY City website, http://diycity.org/. Chris Anderson's "Vanishing Point theory of news" was originally published at http://www.longtail.com /the_long_tail/2007/01/the_vanishing_p.html.

Journalism. The Pothole Paradox

For more on the debate over the future of journalism, see my discussion with Paul Starr published in *Prospect*, "Are We on Track for a Golden Age of Serious Journalism?" For another, more skeptical, overview of journal-ism's future in the age of peer networks, see "Confidence Game," by Dean Starkman, published in the *Columbia Journalism Review*. My book *Where Good Ideas Come From* explores the eco-system metaphor for information technology in more detail; see also James Boyle's "Cultural Environmen-talism and Beyond," published in *Law and Contemporary Problems*. Cass Sunstein's theories on echo chambers appear in his books *Republic.com* and *Echo Chambers: Bush v. Gore, Impeachment, and Beyond*, the latter a digital publication from Princeton University Press. For a rebuttal to Sun-stein's theory, see "Ideological Segregation Online and Offline," by Mat-thew Gentzkow and Jesse M. Shapiro (http://faculty.chicagobooth.edu/ jesse.shapiro/research/echo_chambers.pdf). For more on the power of di-versity, see Scott E. Page's *The Difference: How the Power of Diversity Creates Better Groups, Firms, Schools, and Societies* and his recent *Diversity and Com-plexity*. See also Henry Farrell's "The Internet's Consequences for Poli-tics," http://crookedtimber.org/wp-content/uploads/2011/09/ARPS.pdf.

Technology. What Does the Internet Want?

On the debate over the role of social media in supporting protest movements, see Malcolm Gladwell's *New Yorker* article "Small Change: Why the Revolution Will Not Be Tweeted" (http://www.newyorker.com /reporting/2010/10/04/101004fa_fact_gladwell) and his follow-up blog post "Does Egypt Need Twitter?" (http://www.newyorker.com/online /blogs/newsdesk/2011/02/does-egypt-need-twitter.html); Evgeny Moro-zov's *The Net Delusion: The Dark Side of Internet Freedom*; as well as a thoughtful overview of the debate by Bill Wasik at *Wired*, "Gladwell vs. Shirky: A Year Later, Scoring the Debate over Social-Media Revolutions" (http://www.wired.com/threatlevel/2011/12/gladwell-vs-shirky/all/1). Originally coined by the psychologist James J. Gibson, the concept of "affordances" was developed by the interface theorist and designer Donald Norman in his book *The Design of Everyday Things*. On the visual affordances of television, see Neil Postman's *Amusing Ourselves to Death: Public Discourse in the Age of Show Business*. Few books have thought more richly about the long-term affordances of technology itself than Kevin Kelly's *What Technology Wants*.

Incentives. We Have a Winner!

For more on the Royal Society of Arts, see *A History of the Royal Society of Arts*, by Henry Trueman Wood, and *Joseph Banks and the English Enlightenment: Useful Knowledge and Polite Culture*, by John Gascoigne. On the innovation threat posed by intellectual property restrictions, see Lawrence Lessig's *The Future of Ideas*, and my own *Where Good Ideas Come From*. Ayn Rand's views on patents come from an essay, "Patents and Copyrights," included in the collection *Capitalism: The Unknown Ideal*. John Harrison's story is told in Dava Sobel's popular *Longitude: The True Story of a Lone Genius Who Solved the Greatest Scientific Problem of His Time*. Beth Noveck's inspirational work with crowdsourced patent review, dubbed "peer to patent," is described in her *WikiGovernment*. For more on Jon Schnur and the origin and implementation of the Race to the Top program, see Steve Brill's entertaining *Class Warfare: Inside the Fight to Fix America's Schools*. Interviews with both Noveck and Schnur are included in an anthology I edited, *The Innovator's Cookbook*.

Governance. Liquid Democracies

For more on the history of participatory budgeting, see Rebecca Abers's *Inventing Local Democracy: Grassroots Politics in Brazil*. Lawrence Lessig's critique of the current campaign-finance system, with his proposal for democracy vouchers, appears in his recent book *Republic, Lost: How Money Corrupts Congress—and a Plan to Stop It*. Robert Brooks is quoted from his 1910 book *Corruption in American Politics and Life*. For more on the origins of proxy voting, see Lewis Carroll's "The Principles of Parliamentary Representation." The German Pirate Party has implemented "liquid democracy" techniques with some success in recent years; see "New Politics, Ahoy!" by Steve Kettmann in *The New York Times* of May 1, 2012. For more high-tech implementations of these ideas, see "Toward Delegated Democracy: Vote by Yourself, or Trust Your Network," by Hiroshi Yamakawa, Michiko Yoshida, and Motohiro Tsuchiya (http://www.akademik.unsri .ac.id/download/journal/files/waset/v1-2-19-4.pdf).

Corporations. Conscious Capitalism

For more on the "conscious capitalism" movement, go to consciouscapital-ism.org; John Mackey's essay "Conscious Capitalism Creating a New Paradigm for Business" is available there. The stakeholder approach is explored in *Firms of Endearment: How World-Class Companies Profit from Passion and Purpose*, by Rajendra S. Sisodia, David B. Wolfe, and Jagdish N. Sheth. On the history of Silicon Valley and stock options, see Joseph Blasi, Douglas Kruse, and Aaron Bernstein, *In the Company of Owners: The Truth About Stock Options (and Why Every Employee Should Have Them)*. Tom Wolfe's essay "The Tinkerings of Robert Noyce" appeared in the December 1983 issue of *Esquire*. More information on the Cass Business School study of employee-owned businesses can be found at http://www .cassknowledge.com/inbusiness/feature/employee-owned-businesses. *Wired*'s Tim Carmody published a smart annotation of Zuckerberg's S-1 letter at http://www.wired.com/epicenter/2012/02/facebook-letter -zuckerberg-annotated/.

CONCLUSION. THE PEER SOCIETY

An excellent time line of the SOPA/PIPA history is available at http:// sopastrike.com/timeline.

Index

ABOUT THE AUTHOR

Steven Johnson is the author of the bestsellers *Where Good Ideas Come From*, *The Invention of Air*, *The Ghost Map*, *Everything Bad Is Good for You*, *Mind Wide Open*, *Emergence*, and *Interface Culture*, and is the editor of the anthology *The Innovator's Cookbook*. He is the founder of a variety of influential websites—most recently, outside.in—and writes for *Time*, *Wired*, the *New York Times*, and the *Wall Street Journal*. He lives in Marin County, California, with his wife and three sons.

Steven Johnson writes about big ideas.

The world is changing—faster than ever. There are more big ideas and more good ideas out there. And Steven Johnson tells us both where they came from and where they can take us.

His books are insightful, wide-ranging, about the future and about our history. They are essential for business, innovation, technology, history, and science readers.

Bill Clinton gave a talk recently where he discussed some of Steven Johnson's books:

"There's an interesting book—if you want to be optimistic about the future—by Steven Johnson, who's a great science writer. It's called *Future Perfect*. [Two of his earlier] books, one of them is called *The Ghost Map*, which is about how the cholera epidemic was solved in London; and one's called *The Invention of Air*, which is about the discovery of oxygen."

Steven Johnson's curious, dynamic, creative mind reveals a fascinating world of ideas and innovation.

Steven Johnson has big ideas.

© Nina Subin

T282-0413

Everything Bad Is Good for You
How Today's Popular Culture Is Actually Making Us Smarter

Steven Johnson's hallmark classic on pop culture and technology. In this provocative, unfailingly intelligent, thoroughly researched, and convincing book, Johnson draws from fields as diverse as neuroscience, economics, and media theory to argue that the pop culture we soak in every day—from *The Lord of the Rings* to Grand Theft Auto to *The Simpsons*—is actually sophisticated and, far from rotting our brains, is actually posing new cognitive challenges that are making our minds immeasurably sharper.

"Iconoclastic and captivating."　　　　　　　　　　**—The Boston Globe**

"Persuasive...The old dogs won't be able to rest as easily once they've read *Everything Bad Is Good for You*, Steven Johnson's elegant polemic."　　　　　　　　　**—Walter Kirn, *The New York Times Book Review***

"Wonderfully entertaining. Steven Johnson proposes that what is making us smarter is precisely what we thought was making us dumber: popular culture."　　　　　　　　　　**—Malcolm Gladwell, *The New Yorker***

The Ghost Map: The Story of London's Most Terrifying Epidemic—and How It Changed Science, Cities, and the Modern World

A *New York Times* Notable Book

A riveting page-turner about a real-life historical hero, Dr. John Snow. In the summer of 1854, London has just emerged as one of the first modern cities in the world. But lacking the infrastructure—garbage removal, clean water, sewers—necessary to support its rapidly expanding population, the city has become the perfect breeding ground for a terrifying disease no one knows how to cure. As the cholera outbreak takes hold, a physician and a local curate are spurred to action—and ultimately solve the most pressing medical riddle of their time.

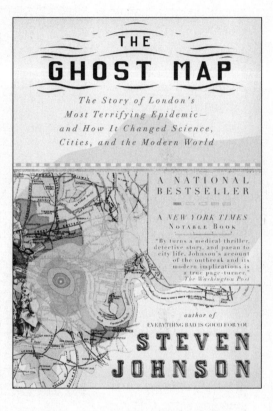

Johnson illuminates the intertwined histories and inter-connectedness of the spread of disease, contagion theory, the rise of cities, and the nature of scientific inquiry.

"Thrilling."   —*GQ*   "Vivid."   —*The New Yorker*

"Marvelous." —*The Wall Street Journal*   "Fascinating."

—*The New York Times Book Review*

The Invention of Air
A Story of Science, Faith, Revolution, and the Birth of America

A book of world-changing ideas wrapped around a compelling narrative, a story of genius and violence and friendship in the midst of sweeping historical change that provokes us to recast our understanding of the Founding Fathers. *The Invention of Air* is the story of Joseph Priestley—scientist and theologian, protégé of Benjamin Franklin, friend of Thomas Jefferson—an eighteenth-century radical thinker who played pivotal roles in the invention of ecosystem science, the discovery of oxygen, the founding of the Unitarian

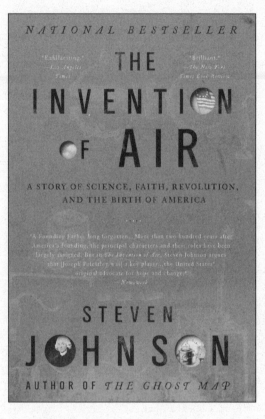

Church, and the intellectual development of the United States. Steven Johnson here uses a dramatic historical story to explore themes that have long engaged him: innovation and the way new ideas emerge and spread, and the environments that foster these breakthroughs.

"Entertaining…clear-sighted and intelligent." —*The New Yorker*

"An exemplar of the postcategorical age…Brilliant." —*The New York Times Book Review*

T287-0413

Where Good Ideas Come From
The Natural History of Innovation

A *New York Times* bestseller • An *Economist* Best Book of the Year

The printing press, the pencil, the flush toilet, the battery—these are all great ideas. But where do they come from? What kind of environment breeds them? What sparks the flash of brilliance? How do we generate the breakthrough technologies that push forward our lives, our society, our culture? Steven Johnson's answers are revelatory as he identifies the seven key patterns behind genuine innovation, and traces them across time and disciplines. From Darwin and Freud to the halls of Google and Apple, Johnson investigates the innovation hubs throughout modern time and pulls out the approaches and commonalities that seem to appear at moments of originality.

STEVEN
JOHNSON

"A FIRST-RATE STORYTELLER."—*THE NEW YORK TIMES*

WHERE GOOD IDEAS
COME FROM

A *NEW YORK TIMES* BESTSELLER

THE NATURAL
HISTORY OF
INNOVATION

"ENTERTAINING AND SMART."
—*LOS ANGELES TIMES*

FROM THE BESTSELLING
AUTHOR OF *EVERYTHING
BAD IS GOOD FOR YOU*
AND *THE INVENTION
OF AIR*

"[A] rich, integrated, and often sparkling book. Mr. Johnson, who knows a thing or two about the history of science, is a first-rate storyteller."

—*The New York Times*

"A vision of innovation and ideas that is resolutely social, dynamic, and material… Fluidly written, entertaining, and smart without being arcane."

—*Los Angeles Times*

The Innovator's Cookbook
Essentials for Inventing What Is Next

Innovation is one of today's buzzwords for a reason. The need to push forward, find new paths and new ideas in an ever-evolving world, is a vital part of business, of education, of politics, of our daily lives. This is an essential book for anyone interested in innovation: the key texts on the topic from a wide range of fields, as well as interviews with successful, real-world innovators, prefaced with an original essay from Johnson that draws upon his own experience as an entrepreneur and author.

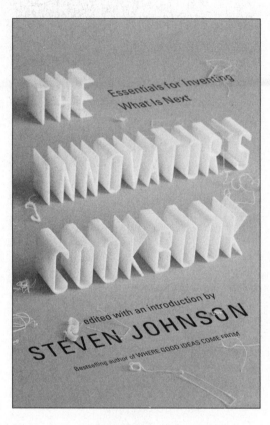

Includes writing from:

Stewart Brand	Teresa Amabile	Beth Noveck
Clayton Christensen	Peter Drucker	Jon Schnur
	Amar Bhidé	Katie Salen
Richard Florida	Ray Ozzie	Brian Eno